# Statistical Literacy

## A GUIDE TO INTERPRETATION

**Dennis G. Haack**   *University of Kentucky*

*Lawrence H. Muhlbaier*
*1979*

D1173285

*DUXBURY PRESS*
*North Scituate, Massachusetts*

*Statistical Literacy: A Guide to Interpretation* was edited and prepared for composition by Jill O'Hagan. Interior design was provided by Dorothy Booth. The cover was designed by Elizabeth Anne Spear.

**Duxbury Press**
A Division of Wadsworth Publishing Company, Inc.

**Library of Congress Cataloging in Publication Data**

Haack, Dennis G., 1944–
    Statistical literacy.

    Includes index.
    1. Statistics.    I.  Title.
QA276.12.H32        001.4'22        78-15381
ISBN 0-87872-183-5

Printed in the United States of America
1 2 3 4 5 6 7 8 9 — 83 82 81 80 79 78

*An undertaking of this magnitude
necessarily involves a family,
rather than just an individual, commitment.
I would therefore like to dedicate this book
to my family:*

*Geri
Mary Clancy
Daniel Song*

# Contents

# Preface

All teachers of statistics courses are doubtless aware of the negative attitude students often have toward learning about statistics. Further, most of us have wondered where this problem of attitude arises — with the teacher, the learner, or the course itself. Many teachers feel that students don't know what's best. Statistics is a bitter pill that must be taken for the student's own well-being. The same teachers often feel that the students will later look back on their elementary statistics course and realize that it was indeed valuable.

But does the problem really lie with the teacher or with the type of initial statistics course we give students? Few who teach statistics courses are data analysts; yet this is the course we try to teach. Using Greek and other mathematical symbols that few students understand we try to teach students data manipulation. In addition, we try to teach students when and how to use techniques such as a $t$-test; yet it is doubtful that any of them will ever have a need to analyze data using such a test, and it is just as doubtful that those who will have to run a $t$-test will do so properly. Is it necessary for a first course in statistics to teach data manipulation, or would we do better to emphasize the interpretation of statistics?

In order to answer this question, let us first decide how students are likely to come in contact with statistics. Rather than designing an experiment and collecting and analyzing the data, students are more likely to read the results of an experiment and need to interpret those results. Some students may never read research reports, but all will be exposed to newspapers, magazines, radio, and television. Through the media everyone comes in contact with statistics each day.

Statistics is more a language in this regard than a research tool. This language is often misused, giving our profession a bad name. Misuse of the language of statistics is *statistical doublespeak*. Why don't we teach students how to understand statistics as a language so they can detect the statistical doublespeak they encounter in the media as well as in their fields of study? This book and the course for which it is designed have this very objective: *to teach students how to interpret statistics so they can detect statistical doublespeak in the media and in their fields of study.*

If we decide that a first course in statistics should emphasize the interpretation of statistics, what adjustments can be made in the usual course? First, symbolic formulas are not as necessary. Students are to be concerned with what a statistic means, not how to calculate the statistic. Many statisticians agree that the amount of formulation in elementary statistics courses needs to be reduced. It is revolutionary, however, to eliminate all formulas. Many feel that this is a drastic step — and it is! But I have found that students accept it enthusiastically.

Elimination of symbolic language is but one feature of this text. Another problem I have encountered is teaching the half-dozen different tests of a particular statistical hypothesis. Students lose sight of the basic philosophy behind testing because they expend their energies trying to choose the "best" test from among the many possible ones. Again, this text makes a sharp departure from the traditional approach. I concentrate on large sample tests of the proportion of a dichotomous population. Indeed, you can easily restrict the discussion of statistical inference to sections involving the proportion. Clearly, such tests are only a tip of the iceberg, but the purpose is to get students to understand the principles of hypothesis testing without trying to make them decide which of many competing tests is "best."

If the symbolism is eliminated and the number of cases of statistical inference reduced, what is put in their place? Many of the topics that are not usually covered in a first course in statistics can be presented. These topics have been chosen in accordance with the course objective: to teach students how to interpret statistics so they can detect statistical doublespeak in the media and in their fields of study. Through the media students will primarily come in contact with sample survey results, economic statistics, and the results of epidemiological research. With the recent surge in epidemiological research into the causes of cancers and heart disease, interpretation of the results of such research is a topic to which all students should be introduced. I also discuss the design of sample surveys. This is an area of much statistical doublespeak in the media. I cannot talk about all of the many economic statistics, but I do discuss the Consumer Price Index in some detail as well as the calculation of constant dollars. Both are topics of everyday importance to students.

This course ideally gets students to critically evaluate statistics in their area of study. Clearly, I cannot discuss in detail all such applications. What I do is present the basics of descriptive statistics, inferential statistics, experimental design, correlation, and regression. With this foundation I have found that students can apply basic principles of interpretation to the specific statistics in their area of study. Of course, if the students are all in one area of study, there isn't as much of a problem. The difficulty comes when a class is made up of students from many different areas. A teacher just is not likely to have expertise in all the subject areas for which there are students. The adventurous are encouraged to venture into all the areas — there is so much for both the teacher and the student to learn. A teacher in this situation can restrict the emphasis to media examples. You'll note that this book has such an emphasis.

What can you, the teacher, look forward to? First, you'll teach a course that isn't likely to become boring to you, since material can be directed to statistics being emphasized in the media. For example, during elections preelection polling can be emphasized. Following the outbreak of Legionnaire's disease my class spent extra time studying epidemiology. The course is consequently fun to teach. Second, your students will become genuinely interested in learning about statistics — What a treat this is!

# Acknowledgments

I would like to take this opportunity to acknowledge some of the assistance I received in the development of the course Statistics: A Force in Human Judgment at the University of Kentucky. This text is an outgrowth of my work to develop this course. The committee, chaired by me, that originally looked into the possibilities of developing a nonsymbolic statistics course was made up of Drs. R. L. Anderson, H. E. McKean, D. Culver, F. Smith, and P. Purdue, all of the Department of Statistics at the University of Kentucky.

Many people have reviewed the different drafts of this manuscript. Their comments were very much appreciated. I'd like to especially acknowledge Carol Beal for her very careful reading of this manuscript. Carol's comments were especially appreciated.

I would also like to thank my typist, Phyllis Renfro, for her diligence and patience throughout this endeavor.

Without the support, encouragement, and help of my wife, Geri, it would have been impossible to complete this project.

My thanks to all of you.

# Disclaimer

Examples selected for discussion in this book do not represent all potential sources. Selection of examples follows personal contact with the example through the media or contact by students or colleagues. If more examples come from one source than from another source, it is a reflection of this mode of contact with potential examples.

The criterion for inclusion in this book was that an educational point can be made from a discussion of the example. I judged the point to be an aid to the reader in evaluating the language of statistics that pervades our media, be it radio, television, newspapers, or newsmagazines.

# Introduction

IT IS USUALLY THOUGHT that statisticians are people who do nothing but manipulate numbers. At the end of a typical sports broadcast a statistician is thanked for keeping track of baseball batting averages, basketball shooting percentages, or football yardage data. Thus, to most of us a statistician's only tasks appear to be those of cataloguing, tabulating, and filing the numbers that pervade our lives. "How boring!" people often think.

No one would deny that statisticians work with numbers, but statistics isn't always the dull discipline that the media imply it is. For instance, statisticians work closely with researchers to design experiments and sample surveys. Together they collect, analyze, and interpret the data generated from the experiments or surveys. Moreover, statisticians are involved in many different kinds of research, thereby getting an overall view of a whole area of study, which the researchers themselves often do not get. For example, a biostatistician can be involved in the many aspects of cancer research, although particular researchers may see only the small area they are investigating.

Statistics is an important aspect of many fields including business, marketing, and medical and educational research. Because of the broad application of statistics, most of you are required to take at least one course in statistics. Unfortunately, your contact with statistics can be a bitter experience. The course is anticipated like the plague, and all too often it is just as expected — a plague.

Why is this? Typically, a statistics course concentrates on data manipulation, which usually involves two bitter pills for you to swallow. First, there are mathematical symbols and formulas. You are expected to become familiar with Greek letters such as $\mu$, $\sum$, $\alpha$, and $X^2$ that accompany so many statistics courses. You must also learn formulas like $s^2 = \sum (X_i - \overline{X})^2/n$. And to add insult to injury, there are many different forms for the same equation. For example, $s^2 = \sum (X_i - \overline{X})^2/n = n^{-1} [\sum X_i^2 - (\sum X_i)^2/n] = \sum X_i^2/n - \overline{X}^2$. In addition, if you try to read other statistics texts the formulas might change. For example, some texts define $s^2$ as $\sum(X_i - \overline{X})^2/n$, and others use the formula $\sum (X_i - \overline{X})^2/(n\text{-}1)$.

After you master this new language, you must try and swallow a second pill: you are expected to learn the many different situations that can occur in data analysis. For example, in testing a hypothesis about the mean of a normal distribution you must consider three cases: variance known; variance unknown with a small sample size; and variance unknown with a large sample size. Then you must look at the

1

nonnormal situations. Finally, to round out the picture, you may have to learn to test hypotheses concerning the proportion, the variance, and the correlation coefficient. As a result of all this, your expectations are fully realized — you feared the worst, and you got it.

The main objective of the statistics course we have just discussed is to teach you how to analyze data. On the other hand, the main objective of this book and the course for which it is designed is to give you statistical literacy. Few of you will go on to become statisticians who will need to know how to collect and analyze your own data. But many of you will do work that will require communication with statisticians, and all of you will undoubtedly need to understand statistics in your daily lives. We cannot watch a news broadcast, read a newspaper or magazine, or listen to the radio or television without being bombarded by statistics. We cannot properly evaluate the performance of our elected officials if we do not understand the statistics on which they claim to base their decisions. And without understanding the *language* of statistics, it is not difficult for us to be manipulated by advertisers who use statistics as a guise for scientific accuracy.

The misuse of statistics by public officials and advertisers is called *statistical doublespeak*. We are all susceptible to misuses of the English language, but we are all particularly vulnerable to the manipulation of statistics. Thus, the objective of this book and this course will be to aid you in detecting statistical doublespeak by helping you to learn the language of statistics.

To learn the language of statistics, we must first learn to distinguish between meaningless numbers and credible statistics. After the introductory discussions of chapter 1, chapter 2 will discuss how to make this distinction. Having determined that a number may indeed be a credible statistic, we proceed with the second phase of our study of the language of statistics: how to interpret a statistic. In chapter 3 we will look at the proper use of statistics to describe a set of numbers or data. Chapters 4 through 6 discuss statistical inference, which is concerned with trying to characterize an entire collection of people when we have studied only a small part of the collection. Chapters 7 through 10 discuss some common uses and techniques of statistics including design of experiments and sample surveys, correlation and regression, and index numbers. Chapter 7 deals specifically with the interpretation of statistics relative to epidemiology, the study of the cause, transmission, and prevention of disease. As before, our objective in these chapters will be to help you to interpret the statistics you encounter through the media now and in the future as well as the statistics you will likely encounter in your work after graduation. In chapter 11 we return to our discussion of doublespeak to reinforce our objective of helping you to evaluate the use and misuse of statistics by public officials and advertisers.

Throughout the text our discussions are in the English language. The obstacles of mathematical symbols and formulas have been omitted. We discuss statistics much as you might discuss any non-mathematical subject. Our emphasis is on understanding and interpreting the statistics you are bound to encounter throughout your career. We hope that you will enjoy and benefit from this approach to statistics.

# The Language of Statistics 1

WE WOULD ALL AGREE that statistics involves numbers. But there is little agreement about just what statistics is and what exactly statisticians do. Statistics is a tool of the researcher. Statisticians work with researchers in a wide variety of fields to design experiments or sample surveys and to collect, analyze, and interpret the data generated. Biostatisticians assist in medical research, and econometricians work in economic research. Statisticians also work with industrial personnel in operations research including quality control, and in stochastic processes, which are involved with problems such as traffic control and the routing of telephone calls. Research in agriculture involves the statistician in the development of better crops, and research in the behavioral sciences involves the statistician in the development of, let's say, better educational methods.

But few people ever come in contact with statistics in a research situation. Most of us have our primary contact with statistics through statistics as a language. We constantly encounter the language of statistics when we read newspapers and newsmagazines or when we listen to news broadcasts on radio and television. It is in the latter sense that we will be considering statistics in this book — that is, we will be looking at statistics as a *language.* The objective of our discussions will be to learn how to understand and "translate" this language.

The language of statistics is often misused. We refer to the misuse of statistics as *statistical doublespeak.* Examples of such doublespeak are everywhere. Advertisers often try to sell a product using statistical statements that imply accuracy through the use of numbers that may or may not be meaningful statistics. Public officials often base their decisions on statistics generated through public opinion polls. But we cannot properly evaluate either advertisements or political decisions if we do not recognize worthless numbers and understand the limitations of even meaningful statistics. We are too often manipulated by advertisers and politicians because we cannot properly evaluate the numbers they use.

We will begin our discussion of the language of statistics by distinguishing between descriptive and inferential statistics. A *statistic* is a number. *Descriptive statistics* are numbers used to describe particular characteristics of a set of data. *Inferential statistics*, on the other hand, are numbers obtained by studying only a part of a collection of numbers in which we are interested. Inferential statistics are used to inductively characterize the collection from which the part was taken. When we use the word *statistics* in the singular, we are referring to the subject of statistics.

The study of statistics is often depersonalized because time is seldom taken to talk about the people who are responsible for the development of the statistical procedures discussed. To put the people back in statistics, we will speak briefly of the history of statistics in this chapter, and we will continue these historical discussions throughout the book.

## Statistical Doublespeak

At the Sixty-first Annual Convention of the National Council of Teachers of English (NCTE) the council resolved to combat the misuse of language by public officials and advertisers. Specifically, the NCTE decided to "find means to study dishonest and inhumane uses of languages and literature by advertisers, to bring offenses to public attention, and to propose classroom techniques for preparing children to cope with commercial propaganda" and to "find means to study the relation of language to public policy, to keep track of, publicize, and combat semantic distortion by public officials, candidates for office, political commentators, and all those who transmit through the mass media."

As a result of those resolutions, in 1972 the NCTE formed a Committee on Public Doublespeak. The word *doublespeak* comes from a combination of two words, "doublethink" and "newspeak," used by George Orwell in his classic book *1984. Webster's New Collegiate Dictionary* 1977 defines doublespeak as "inflated, involved, and often deliberately ambiguous language."

To illustrate the idea of doublespeak, let us look at one of the first annual awards of the Committee on Public Doublespeak presented in the fall of 1974. The award for gobbledygook went to President Richard M. Nixon's press secretary, Ron Ziegler, for his answer to a question concerning whether Watergate tapes were still intact. Ziegler replied,

> I would feel that most of the conversations that took place in those areas of the White House that did have the recording system would in almost their entirety be in existence but the special prosecutor, the court, and, I think, the American people are sufficiently familiar with the recording system to know where the recording devices existed and to know the situation in terms of the recording process but I feel, although the process has not been undertaken yet in preparation of the material to abide by the court decision, really, what the answer to that question is.

An important mode of doublespeak used in advertisements and by our public officials is statistical doublespeak. What can be more "inflated," more "involved," or more "deliberately ambiguous" than statistics based on fallacies or statistics that are not properly understood?

---

*Statistical doublespeak* is the inflated, involved, and often deliberately ambiguous use of numbers.

---

When someone uses a statistic to defend a position, doublespeak might result in three ways.

First, the statistic may, as we will discuss in this book, appear to be quite credible, but it may be misused. For example, in a newspaper editorial of June 1974 it was reported that "Ford administration strategists relied on grossly distorted statistics to scare enough House Republicans into switching sides on the strip mining bill so that the President's veto would be upheld." The editorial also reported, "A West Virginia professor told congressmen that the administration misused his research in concluding that 36,000 jobs might be lost. In fact, he said, the strip mining bill would increase coal industry employment because more deep mines would be opened."

Second, doublespeak might result from the use of a meaningless number. For example, a nationally syndicated columnist stressed a point concerning the plight of Vietnam veterans by stating that 500,000 Vietnam veterans had attempted suicide. It was later found that this statistic was fallacious.

Third, statistical doublespeak can result when we hear or read about credible statistics whose limitations — for example, the source of the data or how the data were obtained — are not known. Examples include unemployment figures, sample survey results, the Consumer Price Index, and advertising claims of significance. If we do not know the source of the statistics or how they were obtained, we simply cannot evaluate them.

Perhaps the most perplexing aspect of statistical doublespeak is that there always seems to be a number to support each side of a controversy. But if we understand statistics, even this might be explained. For example, some of the numbers being used by proponents of one side of an issue might be meaningless. And since the numbers that are credible statistics are limited in the information they convey, they might be incorrectly interpreted by one side or the other. In other words, all three causes of statistical doublespeak may be coming into play.

In considering how to interpret statistics, we will see that the conclusions we can draw from statistics, no matter how carefully the statistics might have been collected, are limited. We may have to consider many statistics as well as other sources of information before getting an accurate picture. For example, no *one* statistic can accurately describe the state of the economy, just as there is no **one** statistic that proves that smoking causes lung cancer. When we look at the economy, the smoking and health question, or any complex issue, a particular statistic is but a part of the overall picture we would like to paint.

## Types of Statistics

> The word *statistic* refers to a number.

Before we can interpret a statistic, we must first know the type of statistic being considered because a particular number can be used for two purposes — to describe something and to make an inference. Thus, statistics (numbers) can be divided into two major types, descriptive and inferential.

> Statistics used to describe certain characteristics of a set of data are referred to as *descriptive statistics.*
>
> *Inferential statistics* are numbers obtained from a part of a set of data. Inferential statistics are used to try to characterize the collection from which the part was selected.

Common descriptive statistics are indexes of central tendency, which are numbers used to describe the "middle" of a set of data. The word "average" is often used when we are speaking of an index of central tendency. We have all heard someone attempt to describe the average or "typical" member of a particular group of people. For example, we might describe the average American as a person who has a high school education, lives in a household containing 2.5 people, and is 32.9 years old.

Most of the statistics reported by the media are inferential statistics. Ronald A. Fisher, a noted twentieth-century statistician, was referring to inferential statistics when he said, "A statistic is a value calculated from an observed sample with a view to characterising the population from which it is drawn."[1] Two ideas that are important in the use of inferential statistics, as can be noted from Fisher's definition, are those of a population and of a sample.

---

The whole, or the entire, collection of objects or numbers in which we are interested is called the *population.*

The part of the population that is investigated is called a *sample.*

---

If we are planning to buy a car, we are interested in certain facts about the population of all cars that are on sale. A politician considering changes in the Social Security system is interested in information about the population of people who receive (or should receive) Social Security. When a television executive contemplates canceling a television program, the opinions of advertisers and viewers should be considered. Unfortunately, all the required facts about these populations are usually not known, and we have, at best, only partial information about them; that is, we have information only about a sample taken from the population.

There are many reasons why we cannot investigate an entire population and must use a sample. We will look at three of these reasons to illustrate the point.

First, economic considerations (time and cost) may dictate that only a sample be investigated. For example, a typewriter assembly plant may receive thousands of parts from many different contractors. These parts must meet certain specifications; if they do not, they will not fit in place or the assembled typewriters will not work properly. In most instances companies cannot afford the time and salaries required to inspect every part entering their plant. Only a small number of the parts, a sample of parts, is inspected, and a decision is made to accept or reject the entire shipment on the basis of the results of tests run on the sample.

Other situations in which samples are used because of economic considerations include the many types of sample surveys. For example, in 1940 the Census Bureau began to sample the United States population when the cost of getting the answers to more and more questions from more and more people began to exceed available resources. Today a *census,* a complete enumeration or count, of the population furnishes but a small part of the statistical information collected by the Census Bureau. Most data are collected through extensive sampling of the populace.

The second reason for using a sample is that there are times when a whole population cannot be studied because at the time of the investigation the population of interest does not exist. An example of this situation is preelection polling. The population of interest is the collection of all people who will vote on election day. But at the time of the poll this population does not yet exist. At the time of a preelection poll some people have not yet decided for whom they will vote, and others may not even vote. Thus, preelection polls may be fairly accurate measures of public preference at the time the poll is taken but not necessarily of actual voter action on election day.

Let us look at an example. Two months prior to the 1972 New Hampshire presidential primary, a **Boston Globe** poll showed Senator Edmund Muskie ahead of Senator George McGovern by 65% to 18%. However, a poll run two months before a primary is often merely a public recognition survey because one candidate is likely to be better known than another. In this particular primary Muskie, who is from Maine, was better known than McGovern, a South Dakota senator. The **Globe** poll reflected this fact by the large gap between these candidates. But people's opinions change as they get to know the candidates. Also, people who say they intend to vote often end up not voting. This is especially true in primary elections in which voter turnout is likely to be low. The sample on which the **Globe** based its figures was probably not representative of the people who actually voted in the primary. In fact, in the primary Muskie beat McGovern only by a margin of 46% to 37%, which caused Muskie's campaign a serious setback since he was expected to take at least 50% of the vote. Senator Muskie's campaign did not recover from this setback. (For more discussion of improper interpretation of survey results see **Lies, Damn Lies, and Statistics** by Michael Wheeler.[2])

A third reason for using a sample occurs when testing the durability of a product requires destruction of the product. Manufacturers usually test their products before marketing them. Testing the life of a light bulb, for example, requires using it until it burns out. Quite clearly, it would not be feasible for a company to test the entire population of light bulbs before making a marketing or advertising decision.

We see from these examples that in many situations an entire population cannot be investigated and a sample from the population must be investigated instead. In these situations we must use statistical inference and inductive reasoning to draw conclusions about a population when we have studied only a sample from the population. But as you might expect, there is some uncertainty involved with this type of reasoning. Fortunately, the techniques of statistical inference provide us with a method for measuring this uncertainty, namely, probability.

> *Probability* is a measure of the uncertainty or reliability of the conclusion drawn about a population from an analysis of a sample.

Any discussion of inferential statistics necessarily involves us in a discussion of probability. The adjective "inferential" will indicate whether probability must be part of our thought processes in interpreting a statistic.

As we consider examples of the different types of statistics, you should begin to be able to determine how a statistic is being used — that is, you should be able to tell when a statistic is being used to infer from a sample (part) to a population (the whole).

## A History of Statistics

Because most presentations of statistics deal primarily with numbers, it is easy to forget that the body of knowledge in statistics is the end product of the contributions of many people whose personalities as well as the times in which they lived greatly affected their works. To compensate for the oversight, we present a brief overview of the history of statistics. Throughout the book we will refer to the history of the statistics discussed.

The term "statistik" was probably first used by the German philosopher Gottfried Achenwall in the middle of the eighteenth century. The word "statistik" referred to "inquiries respecting the Population, the Political Circumstances, the Productions of a Country, and other Matters of State."[3] While the science of statistics was being studied in Germany, the words "statistics" and "statistical" were introduced into the English language around 1787 by Ebesherd A. W. von Zimmerman (1743–1815). However, the methods of statistics were in use much earlier than that.

The earliest form of statistics was natural, or census, statistics. One of the early attempts at census taking dates back to the ancient Egyptians, around 3050 BC, and census taking may even go back as far as 4500 BC in Babylonia.[4] The first censuses were conducted for the purpose of levying taxes. Beginning with the Roman emperor Augustus (27 BC–AD 17), a population census was conducted in each newly conquered territory. The Romans kept records of births, deaths, and crops as well as an enumeration of the population. William the Norman (William the Conqueror) carried out exhaustive surveys after his conquest of England in 1066. Doomsday lists contained the results of a land census that detailed estate and livestock sizes for taxing purposes.[5]

During the fourteenth century, signs of the approaching marriage of probability and statistics appeared in the insurance business. Marine insurance rates were set using data that had been gathered concerning the success of the transportation of goods. The probability of the successful completion of a shipment was approximated by using the data. For example, insurance for shipping cargo was set from about 12% to 15% of the value of the cargo during the early 1300s. In comparison, overland rates were set from near 6% to 8%.

Then, in 1662 John Graunt (1620–1674) collected data for the setting of life insurance premiums. His work with the **London Bills of Mortality** can be considered as the beginning of acturial science, which is the statistical branch of the insurance business. He observed, for example, that male births exceeded female births but that more men than women died from violence.[6]

It was during the early seventeenth century that inferential statistics began to emerge through the development of probability theory. James Bernoulli (1654–1705), who turned to the study of astronomy and mathematics after receiving a degree in theology, made one of the early important contributions to this development. In his best known work, **Ars Conjectandi,** Bernoulli made one of the first contributions to the theory of relative frequency probability when he proved that as the number of observations increased, "the ratio of observed successful to unsuccessful occurrences will differ from the true ratio within certain small limits."[7] Also, because of this man's efforts we have the phrase "Bernoulli trial," which refers to a trial (experiment) whose outcome is dichotomous (two distinct groups) — a trial in which the outcome is either "success" or "not success." These ideas have proven to be very important in both probability and statistics.

The extensive use in statistics of normal probability models came about from the early work of Abraham de Moivre (1667–1754). As a French refugee in England, de Moivre was unable to get a university appointment although he was known as a very competent mathematician. Consequently, he spent much time consulting with gamblers and working out interesting problems in probability theory. De Moivre was the first to suggest that the sampling distribution of the proportion of successes in a series of independent Bernoulli trials approached a normal distribution (bell-shaped curve) as the number of trials increased.

Among the many other contributors to the development of probability theory were Pierre-Simon de Laplace (1749–1827) and Carl Friedrich Gauss (1777–1855). Both made significant contributions to the theory of least squares. For the prediction problem least squares theory refers to methods of predicting a quantity so that the sum of squared differences between predicted values and actual values is minimal. Gauss is in addition singled out for his work on the distribution of errors, having proposed that errors in measurement are symmetrically distributed. His work added to the extensive use of the normal or Gaussian probability model in statistical inference.

With the development of probability theory from the 1600s through the 1800s, modern inferential statistics required only the application of these statistical "tools" to experimentation. Karl Pearson (1857–1936) made significant contributions in this regard. A prestigious mathematician, Pearson became interested in the application of statistical theory to biological and sociological data. Pearson's many contributions included work with correlation. Pearson was also the cofounder of the first statistics journal, *Biometrika,* in 1901.

The use of statistics in many varied fields of investigation can be attributed to Lambert Adolphe Jacques Quetelet (1796–1874) and Sir Francis Galton (1822–1911). Born in Belgium, Quetelet worked with the application of statistical methods to moral and social phenomena. Originally trained in the arts, he became acquainted with statistics through contact with French masters. "He immediately thought of applying them [statistical methods] to the measurement of a human body, a topic he had become curious about as a painter."[8] Quetelet applied statistical methods to the analysis of data concerning mortality, crime, physical traits, and behavioral and intellectual qualities.

Galton, the son of a prosperous Quaker businessman, was influenced by the work of Charles Darwin. He attempted to demonstrate statistically the inheritance of intellectual and behavioral qualities for which Darwin labelled him a "passionate statistician, preeminently a lover of statistics."[9] Galton's application of statistical methods to the data from many varied areas came from his feelings that "until the phenomena of any branch of knowledge has been submitted to measurement and number it cannot assume the dignity of a science."[10] Galton was responsible for some of the earliest work in regression. He observed that the average height of the offspring of unusually tall people regressed (moved backward) toward an overall average height.

By the early 1900s statistical theory had been developed and was being applied in many areas of investigation. Sir Ronald A. Fisher (1890–1962) formally brought together statistics and the researcher with the 1925 publication of *Statistical Methods for Research Workers.*[11] Fisher made a profound impact on statistical methods, statistical inference, experimental design, and the foundation of statistics. Fisher's work includes contributions to the theories of estimation, the distribution of inferential statistics, experimental design, and the analysis of variance.

George W. Snedecor (1881–1974) pioneered the development and utilization of statistical applications in biological and agricultural experimentation. Snedecor set up the first statistics laboratory in the United States in 1931 at Iowa State University. His Book *Statistical Methods,* which has sold over 100,000 copies, has contributed greatly to the application of statistical methods to experimentation.[12]

The development of decision theory is attributed to the works of Abraham Wald (1902–1950), Jerzy Neyman, and Egon S. Pearson (the son of Karl Pearson). Wald was the first to see the connection between the mathematical theory of games and statistical theory. His work led to the development of statistical decision theory. The test of a statistical hypothesis is an important example of a statistical decision problem. Neyman and Pearson, both of whom are still living, developed the basic theory for hypothesis testing and interval estimation.

We see, then, that probability and statistics have a long history. And the search for new applications, new theories, and new ideas is by no means over.

## Summary

Statistical doublespeak is the inflated, involved, and often deliberately ambiguous use of numbers. Public officials, advertisers, and almost everyone can misuse statistics and cause doublespeak. Statistical doublespeak occurs when the limitations of a meaningful statistic are not known or when a meaningless number is passed off as a meaningful statistic. Only proper understanding of the language of statistics can help us to avoid statistical doublespeak.

To understand the language of statistics we must first be able to decide whether a number is a credible statistic. Then, having made this determination, we must be able to properly interpret the statistic. An important part of proper interpretation involves knowing the limitations of the statistic because each statistic conveys but part of an overall picture we would like to paint about a subject.

We distinguished between statistics that are used to describe a set of data, descriptive statistics, and statistics that are used to infer from a sample to the population from which the sample was drawn, inferential statistics. Being able to make this distinction is important. A descriptive statistic describes a specific aspect of a data set. The only uncertainty involved lies in knowing which descriptive statistic is proper to use in a given situation. Inferential statistics, on the other hand, have a degree of uncertainty that is inherent in proper interpretation of the statistic itself. This uncertainty will be measured by probability. Statements made concerning inferential statistics will be probability statements, not statements of fact.

We must become accustomed to distinguishing between statistics that are used to describe and those used to make statistical inference. Using the phrases "descriptive statistics" and "inferential statistics" will aid us in interpreting statistics in general as different thought processes are needed to understand these two very different types of statistics. The word "inferential" will be a warning that a mode of thought, which is unfamiliar to us, is required for proper interpretation. This mode of thought is tied to probability.

In the following chapter we will talk about certain preliminary considerations that will aid us in distinguishing between meaningless numbers and meaningful statistics. Our remaining discussions will be concerned with proper interpretation of the statistics, whether descriptive or inferential, that we are likely to encounter.

# No
# Comment

Have you ever wondered why we might be "afraid" of statistics? Could it be that the media reminds us daily that our greatest fear is to become a statistic? In the No Comment section of our discussions quotes will be presented from the media. Comment should not be needed. As we read these quotes, we should ask ourselves why we might be reluctant to even think about statistics.

■ *"I never was one to rely much on statistics," a former attorney general of Kentucky said, "You can make them say what you want."*

■ *"Statistics tell us that over 12 million people in this country suffer disabling injuries every year — at home, at work, or in their cars. Another 1,700,000 are disabled by heart conditions, T.B., cancer, arthritis, and rheumatism. If you become one of these statistics, ..."*
*(From an ad for Creditors Disability Insurance)*

■ *"The written or spoken word may be suspect. But numbers, by their very nature, give the impression of great accuracy. Statistics are used to prove almost anything."*
*(Newsmagazine editorial)*

■ *"The Commerce Department reports with some glee that sales and income figures show an easing up of the rate at which business is easing off. This is taken as proof of Mr. Ford's contention that there is a slowup of the slowdown.*

*"It should be noted that a slowup of the slowdown is not as good as an upturn of the down curve, but it is a good deal better than either a speedup of the slowdown or a deepening of the down curve. And it does suggest that the climate is about right for an adjustment of the readjustment.*

*"Turning to unemployment, we find a definite decrease in the rate of increase, which clearly shows that there is a letup of the letdown. Of course, if the slowdown should speed up, the decrease in the rate of increase of unemployment would turn into an increase in the rate of decrease of employment. In other words, the deceleration would be accelerated.*

*"It is hard to tell, before a slowdown slows up, whether a particular pickup will be fast. At any rate, the climate is right for a pickup this season, especially if you are unmarried and driving a convertible . . . but, perhaps we are letting our mind drift away from our work."*

*(Anonymous, obtained from Norman Cotton, Palo Alto, California)*

## Exercises

1.  Find examples of public doublespeak, statistical and verbal, in advertisements or in statements of public officials. Look in newspapers and newsmagazines, and listen to television and radio broadcasts. Keep the statistical examples for later reference.

2.  Consider the advertising claim that a mouthwash kills millions of germs. Is there something that is not being said? Is this an example of statistical doublespeak?

3.  Try to classify statistics found in the news media as descriptive or inferential.

## For Discussion

For Discussion exercises are meant only to stimulate discussion. Do not look for a "right answer" as there may not be one. Feelings about these topics are to a large extent subjective or personal.

1.  There are an increasing number of instances in which major opinion polls such as the Gallup and Harris polls show different results. Have there been any recent examples? How might this happen?

2.  Many politicians gauge public opinion by mailing question-
    naires to constituents who have taken time to write to them.
    Some members of the House of Representatives may even send a
    questionnaire to all registered voters in their district. What are
    the advantages and disadvantages of this method of surveying
    opinion?

3.  Look at the results of a survey you have recently encountered in
    the media. Can you interpret the results? What information do
    you need for proper interpretation? Is this information available
    in the media release?

# Further Readings

If our brief discussion of public doublespeak has inspired interest in this
subject, the following are recommended:

Hugh Rank's *Language and Public Policy* (1974) and Daniel Dieterich's
*Teaching Public Doublespeak* (1975). Both of these books are put out by the
National Council of Teachers of English, Urbana, Illinois. Hugh Rank was the
first chairperson of NCTE's Committee on Public Doublespeak; Dan Dieterich
is the committee's second chairperson.

For a discussion of the use and abuse of the language of science see Irwin
Bross's *Scientific Strategies in Human Affairs: To Tell the Truth* (Hicksville, N.Y.:
Exposition Press, 1975).

If you would like to know what statisticians do and how statistics are
applied to many different areas of investigation, see J. M. Tanur's *Statistics: A
Guide to the Unknown* (San Francisco: Holden-Day, 1972). This is an excellent
collection of readable essays about statistics. In this anthology you are likely to
find an article about the use of statistics in your area of study. If not, you're
sure to find an essay that discusses the application of statistics in an area
you'll find interesting. Look it over.

More information on the history of statistics and probability can be
found in the article by A. L. Dudycha and L. W. Dudycha listed in the footnote
section of this chapter. Another source is the book by E. Pearson and M. G.
Kendall, *Studies in the History of Statistics and Probability* (London: Griffin,
1970).

# Notes

1.  R. A. Fisher, *Statistical Methods for Research Workers* (Edinburgh:
    Oliver and Boyd, 1925), p. 44.
2.  M. Wheeler, *Lies, Damn Lies, and Statistics* (New York: Liveright, 1976),
    pp. 16–17.
3.  G. U. Yule, "The Introduction of the Words 'Statistics,' 'Statistical' into
    the English Language," *Journal of the Royal Statistical Society*[68] (1905):
    392.
4.  H. H. Wolfenden, *Population Statistics and Their Compilation* (Chicago:
    University of Chicago Press, 1954), p. 4.

5.  L. E. Maistrov, *Probability Theory: A Historical Sketch* (New York: Academic Press,1974), p. 4.

6.  J. A. Ingram, *Introductory Statistics* (Menlo Park, Calif.: Cummings, 1974), p. 2.

7.  K. Pearson, "James Bernoulli's Theorem," *Biometrika* 17 (1925): 201–202.

8.  D. Landau and P. F. Lagarsfeld, "Quetelet, Adolphe," *International Encyclopedia of the Social Sciences* 13 (1968): p. 248.

9.  A. L. Dudycha and L. W. Dudycha, "Behavioral Statistics: An Historical Perspective," *Statistical Issues: A Reader for the Behavior Sciences,* ed., R. E. Kirk (Monterey, Calif.: Brooks/Cole, 1972), p. 17.

10.  *Ibid.,* p. 18.

11.  R. A. Fisher, *Statistical Methods for Research Workers* (Edinburgh: Oliver and Boyd, 1925).

12.  G. W. Snedecor, *Statistical Methods* (Ames: Iowa State University Press, 1931).

# Background Influences on Statistics

2

THE FIRST STEP IN PROPER INTERPRETATION of statistics is to **think.** If we would scrutinize statistical statements as carefully as we sometimes scrutinize verbal and written statements, we could detect much of the statistical doublespeak we encounter. The unhappy truth is that we are inclined to give numbers an undeserved aura of authenticity. We too often accept a statistical statement as fact without so much as a question of clarification. The fact that statistics are not scrutinized is the single most important reason for the misuse of statistics.

In this chapter we will consider certain preliminary determinations that are required for proper interpretation of statistics. These background considerations will give us an idea of where to begin if we desire to critically evaluate a statistic. And they will also help us distinguish between meaningless numbers and the meaningful statistics that warrant consideration. Meaningful statistics will be discussed further in the remainder of this book.

The background influences that we will consider to evaluate a statistic are the source, the type of data, definition and measurement problems, and certain considerations concerning the sample survey statistic. The first thing we should ask ourselves about any statistic is its source — that is, where did the number come from? Having scrutinized the source, we must consider the type of data on which the statistic is based. Our interpretation of a statistic must reflect the type of data collected, since certain descriptive and inferential statistics are more appropriate than others for particular types of data.

How certain characteristics are defined and then measured is another important background influence. The definition and measurement problem may have a greater influence on a statistic that comes from the behavioral sciences than on a statistic that comes from agriculture, but this influence must always be considered. For example, it is far more difficult to define and measure intelligence than to define and measure the yield of a crop.

In the case of a sample survey statistic we will look at certain background influences that will affect our interpretation of the statistic. Besides considering the source of the survey data, we will look at the population sampled, the response rate, the method of contact of respondents, the timing of the survey, and the phrasing of questions. Any one of these background influences can signal that a number is meaningless and needs no further consideration.

## The Source

There is a tendency to blame the statistician when a statistic does not say what we would like. Some researchers may do their own statistical work if a statistician does not tell them what they want to hear.

It is especially important to political candidates that statistics put them in a favorable light during an election campaign. A strong desire exists to be ahead in the polls during a campaign since political contributions appear to be tied to a candidate's showing in the polls.[1] In the 1970 senatorial primary in New York incumbent Charles Goodell was shown to be running a poor third in a straw poll run by the *New York Daily News*. An aide to the incumbent claimed that he was contacted by a polling firm and asked if there was interest in a poll that would show Goodell ahead.[2]

A logical first step, then, in the interpretation of a statistic is to determine the source and consider its reliability. Sometimes a source cannot be identified, and the statistic must be rejected. Recall in chapter 1 in the section on statistical doublespeak we mentioned that a nationally syndicated columnist had indicated that 500,000 Vietnam veterans had attempted suicide. In a later column the columnist confessed that he was unable to trace the source of the figure. The statistic had been taken from an article in *Penthouse* magazine. The author of that article had gotten the figure from a publication of a defunct veterans group. That organization's former director said the statistic was obtained from the Council of the Churches of the United States of America, a group that disavowed any knowledge of the figure.

Sometimes a statistic is fabricated. Consider the claim that each year drug addicts steal $4.4 billion worth of goods in New York. When someone inquired into the origin of this figure, it was found that an estimated 100,000 addicts was multiplied by an estimated average daily habit of $30 to get an approximate collective need of $1.1 billion per year. Reasoning that addicts can fence stolen goods for about one quarter of their value, the $1.1 billion figure was multiplied by 4 to get $4.4 billion; yet this figure is greater than the estimated value of *all* stolen goods in New York.

The point should be clear: When you consider a statistic, first ask, "Where did the number come from?"

Our work in tracing the source of a statistic is compounded as there may be more than one source. As in the veterans example given, there may be many secondary sources as well as a primary source. The further back that a statistic must be traced, the more difficult it is to find the primary source, that is, to find the origin of the statistic. The more removed the use of a statistic gets from the primary source, the more skeptical we must be of the statistic. It is not true that the more often a statistic is quoted the more credible it is. We will discuss the sources of some of the statistics we often encounter.

### Government Statistics

We begin with the greatest disseminator of statistics in the world, the United States government. Government officials use statistical information to make decisions. Due to the effect such decisions have on our lives, we should understand these statistics. The way to begin is to consider the reliability of the reporting agencies.

Let us consider the reporting of a change in the unemployment figures. These government statistics are often accompanied by an interpretation of what the change means. We should ask whether there are any political influences on these interpretations.

To keep such influences at a minimum, we would hope that a professional statistician who is concerned about his or her professional integrity would be responsible for reporting the information and answering questions on interpretation. This was, in fact, the case for unemployment and employment figures that are released monthly by the Bureau of Labor Statistics — at least, it was the case until March 19, 1971. On this date the monthly news conferences conducted by a professional statistical staff were abruptly ended.

The concern of statisticians that federal statistical agencies had become politicized was heightened by reassignment of personnel and reorganization in the Bureau of Labor Statistics, the Census Bureau, and statistical agencies in the Commerce Department as well as by the discontinuance of the Urban Employment Survey, which had been gathering information about residents in poverty areas.[3]

The outcry against political influences on federal statistical agencies was heard from professional statisticians and politicians alike. Carl Albert, the then Speaker of the House, said, "The muzzling of these officials who for 20 years now — under Democratic and Republican administrations alike — have provided impartial interpretations of these statistics for newsmen is a most blatant attempt to impede the flow of government information to the public."[4] And statistician Philip M. Hauser said, "On the basis of my own experience with government statistics I know of no administration in which some zealous politician or political 'eager beaver' did not, at some point, try to impair the integrity of statistical reports; but never have I witnessed as widespread and insistent efforts to politicize the statistical enterprise."[5]

The concerns of professional statisticians culminated in the formation of an American Statistical Association — Federal Statistics Users Conference Committee on the Integrity of Federal Statistics. The primary finding of the committee was that there was no evidence that statistics had been altered to support a particular point of view but the independence of federal statistical personnel had been reduced — that is, the statistics appear credible, but the accompanying interpretations must be viewed critically. Since the interpretation of government statistics reported by the news media is heavily influenced by the interpretation given by the reporting agency, media interpretations must likewise be scrutinized.

Statistical doublespeak results when a credible statistic is misused or misunderstood. Since federal statistics appear to be credible, we must know their limitations and be on the guard against misrepresentation of government statistics by public officials.

## Consumer Statistics

Besides government statistics, consumer statistics are quite common. We are bombarded with consumer information through advertising, consumer magazines, and the media. Here too, we must ask about the source of the information.

One source of consumer information is *Good Housekeeping* magazine. *Good Housekeeping*'s Seal of Approval indicates that (1) the product is advertised in *Good Housekeeping*; (2) the magazine is satisfied that the product is good and that claims explicitly made are truthful; and (3) if the product is defective, the magazine will guarantee a refund or replacement.

The *Good Housekeeping* Seal of Approval does not mean that the product was tested or compared to competing products. There is therefore no indication that the product must meet any standards, federal or otherwise.[6] As you can see, it is important for us to know just what the Seal of Approval means before we place our confidence in its rating of a product.

One of the more popular sources of consumer information is the Consumers Union of the United States (CU), publishers of *Consumer Reports* magazine. CU is a nonprofit organization with a staff of about fifty, which is supported through the nearly 2 million subscriptions of *Consumer Reports*.

The information published in *Consumer Reports* is of two types: information on CU testing of new products and questionnaire data obtained from subscribers that give information on the reliability of products. These two types of information come from two different sources and must therefore be interpreted separately.

What is disconcerting about CU as a source of consumer information is a general lack of details on experiment design, data collection, and data analysis that makes it difficult for us to make our own interpretation of the data they collect. We must accept the ratings determined by CU without question.

Another source of consumer information is Underwriters Laboratories, which is prominent in the area of product safety. The UL seal of Underwriters Laboratories appears on the back of many electrical appliances. UL tests a product only upon request of the manufacturer. UL standards cannot be set too high or manufacturers would not submit their products for testing.

To interpret the information disseminated by UL, it is important to note that the UL seal means that the electrical component of the product, not the entire product, has been approved. In light of the preceding discussion, what can be said about a product which carries a UL seal? Equally pertinent is the question of how the absence of a UL seal should be interpreted. It is estimated that the UL rejects 50% of the products it tests.[7] Products known by the manufacturer to be unsafe are not even submitted to UL for testing. We must therefore be suspicious of electrical appliances that do not carry the UL seal.

We see how important it is to consider the source of statistical information. Whether or not the statistical information is consumer oriented, the beginning of a proper interpretation requires a consideration of the source of information.

## Types of Data

Consideration of the type of data upon which a descriptive or inferential statistic is based is important to proper interpretation of a statistic. In every case a *datum* (plural, *data*) is a measurement taken on a population member. We will discuss the various types of data that are commonly used.

### Nominal Data

A measurement system that consists of clearly defined and non-intersecting categories produces a type of data called *nominal*, or *categorical, data*. Each category is given a *name* (*nom* in French). The most common use of categorical data is the categorizing or classifying of responses to opinion surveys. For example, a person's feelings about the president's performance in office may be categorized as "approval," "disapproval," or "no opinion." A preelection survey might categorize responses as "for Candidate A," "for Candidate B," or "undecided." Census surveys are important sources of categorical data. Here categories might concern race, ethnic origin, religion, house ownership, sex, etc.

### Ordinal Data

Categorical data whose categories are ranked according to some criterion are called *ordinal data*. For example, a student might classify a teacher's performance as "well above average," "above average," "average," "below average," or "well below average," just as a teacher classifies a student by assigning a grade of *A, B, C, D, or F* (failure). In both instances the categories could be given numerical codes 5, 4, 3, 2, and 1, respectively, for which a 5 rating is higher than a 4, a 3 is higher than a 1, etc. However, aside from *ordering* a student's opinion of a teacher, or vice versa, the numerical codes have no meaning — that is, a 5 does not mean 5 units of something; 5 merely means "more than 4." Moreover, considerable subjectivity enters into the assignment of codes.

Other sources of ordinal data include consumer preference surveys. In such a survey a person might be asked to compare, let's say, three different products by ranking them 1, 2, or 3.

### Interval Data

*Interval data* are data that consist of actual measurements for which differences between each pair of data are meaningful. For example, if maximum temperatures are recorded each day in Chicago, a 3 July reading of 33°C can be compared to a 7 December high of 6°C by noting a 27° difference. This difference is meaningful in itself. For example, two days with 29° and 2° maxima would also have a 27° difference. Contrast this with ordinal differences between, let's say, 5 and 4 as compared to 2 and 1. The common difference of 1 in the arbitrary ordered scale does not create any meaningful mental image, and, in fact, has a totally different meaning from a measured arithmetic difference. Note, however, that in the temperature example, the zero is arbitrary on the Celsius scale in the sense that 4° is not 4 times as hot as 1°. Only differences are meaningful.

*Ratio data* are interval data for which ratios are meaningful; that is, ratio data involve a scale of measurement that has a defined zero. For example, an 8 lb newborn baby really is twice as heavy as a 4 lb newborn; a 200 lb person is really twice as heavy as a 100 lb person. Yet note that differences here are also meaningful; the newborn babies are 4 lbs apart, just as 10 lb and 14 lb babies are; a 200 lb man is dwarfed by a 300 lb football player by the same margin that he exceeds his 100 lb wife, though the ratios are not the same.

### Continuous and Discrete Data

*Continuous data* are interval or ratio measurements that can theoretically vary by arbitrarily small amounts, and *discrete data* are interval or ratio data that are not continuous. Observations such as length, weight, temperature, and pressure are continuous in the sense that it is always theoretically possible to get an observation between any other two observations, no matter how close they may be. Thus, if we have two men who weigh 183.25 lb and 183.26 lb, respectively, it is theoretically possible to find other men with weights between 183.25 lb and 183.26 lb.

The best illustration of discrete data is *count data.* The number of hits made by major league baseball players in a given time period is a discrete data set (set of data) since the data are always whole numbers (i.e., 0, 1, 2, 3, and so on). Note that this is ratio data since zero is meaningful. Shoe size is an example of discrete interval data that are not count data but may serve as an approximation of foot length, which is an example of continuous (ratio) data.

It is important to know the type of data on which a statistic is based. Some descriptive statistics and statistical inferences are appropriate for one type of data but are not appropriate for other types of data.

# Definition and Measurement Considerations

Our purpose in this section is to emphasize that proper interpretation of statistics is impossible if we do not know how the characteristics being measured are defined. The importance of knowing what definitions are being used can be illustrated with two examples. We will see how contradictory statistics, a form of statistical doublespeak, may be the result of different people using different definitions.

The first example has to do with the variety of statistics that might be used to compare the populations of the cities of New York and London. Knowing what areas were used to measure population is essential to interpreting the statistics. The information is taken from the year 1955.

Whether the city with the world's greatest population is New York or London depends on what areas are referred to by "New York" and "London." The City of London proper had a population in 1955 of only about 5,200,000 and New York County, or Manhattan, one of the five boroughs of New York City, had 1,910,000. The analogous political units, however, are the City of New York, with a population 8,050,000 in 1955, and the County of London, 3,325,000 in 1955. Each of these is a municipality made up of boroughs, 29 in London and 5 in New York.

A comparison often made (though inaccurately) is that between Greater London and the City of New York — probably because of the coincidence that the City of New York when it was formed by consolidation of New York, Brooklyn, and other areas in 1898, was referred to as "Greater New York." "Greater London" with a 1955 population of 8,315,000 is defined as the area within 15 miles of the center of the City of London. It has been estimated that the area within 15 miles of the center of New York has a population of 10,350,000.

The "New York Standard Metropolitan Statistical Area" however, had a 1955 population of 13,630,000. A metropolitan area defined for London on a basis similar to that used for New York would have a population of approximately 10,000,000.[8]

A Standard Metropolitan Statistic Area (SMSA) is defined by the United States Bureau of the Census as counties with one county containing a central city. The number of counties in a SMSA is determined according to certain socioeconomic criteria (i.e., 20% of the labor force of a county commutes to the central city). A central city is either a city of 50,000 or more or a combination of cities that satisfy certain specific criteria, which need not be discussed here.

The second example illustrating the importance of definitions involves unemployment statistics. Different definitions of unemployment by reputable agencies probably caused the reporting irregularities shown in table 1. Proper interpretation of any of the figures in table 1 would require an investigation of how each agency defined unemployment.

A president's Committee to Appraise Employment and Unemployment Statistics (the Gordon Committee) took an early important

### TABLE 1   Unemployment Estimates for November 1935

| Agency | Unemployment Figures |
|---|---|
| National Industrial Conference Board | 9,177,000 |
| Government Committee on Economic Security | 10,915,000 |
| American Federation of Labor | 11,672,187 |
| National Research League | 14,175,000 |
| Labor Research Association | 17,029,000 |

Source J. B. Cohen, "The Misuse of Statistics," Journal of the American Statistical Association 33 (1935): 664.

look into the problem of defining employment and unemployment. Its report of 1962, *Measuring Employment and Unemployment*, set forth recommendations for changes in the definition of unemployment. The committee's recommendations are the basis for the definitions that are presently used.

Unemployed persons are people who are not working but have made an attempt to find a job within the previous 4 weeks.[9] Persons who are on layoff or waiting to report to a job are also classified as unemployed. We observe that a person who is out of work but is not trying to find work is not considered to be in the labor force and therefore is not counted.

Employed persons are people who did any work at all as paid employees during the survey week.[10] People who are temporarily absent from work because of illness, bad weather, vacation, a labor dispute, or for personal reasons are classified as employed.

In statistics we seek what is called a *workable definition*. To a statistician a definition of a trait or a characteristic is workable if credible measurements can result.

One of the most difficult searches for a workable definition involves the term "intelligence." Some early attempts at defining intelligence include: the ability to do abstract thinking; the power to respond well; the property to act better in novel situations; the power of autocriticism; and the ability to act purposefully.[11]

These definitions are not workable; that is, it would be difficult, if not impossible, to measure intelligence defined in any of the above ways. A workable definition of intelligence has not yet been found. To date the only workable definition of intelligence would be "that which intelligence tests test."[12]

For many psychological characteristics such as aptitude, achievement, and intelligence it is quite difficult to develop workable definitions. This difficulty should influence our interpretation of any statistical analyses based on measurements of psychological characteristics.

Now let us consider some specific measurement problems. The Graduate Record Examination (GRE) is a standardized test designed to help predict success in graduate school. Many statistical investigations have been carried out to determine how well this particular test works. But how do we define and measure graduate school success? The quality of the GRE cannot be accurately judged unless we know what success is.

There have been many attempts to define success.[13] Grades are one way. Grades are readily available and are indicative of a student's achievement. Yet many faculty doubt that grades are a good way to describe the more important outcomes of graduate education. Comprehensive exams are often used to measure graduate school success. But a circularity exists when test scores (GRE scores) are used to predict test scores (comprehensive exam scores). Faculty evaluation and whether a student gets a degree are other success criteria. Faculty ratings can be arbitrary, and degree attainment is as much a measure of persistence as an indicator of success.

We could argue for and against all of these success criteria. Yet when we want to measure success in school, some combination of these 4 criteria is usually used. Definition and measurement problems in the behavioral sciences are quite profound, and we have to be aware of this when interpreting statistics in this area.

Definition and measurement problems also exist for nominal data. Categories must be carefully defined. The population observations must be properly categorized. This is not necessarily an easy task. Consider, for example, the classification of tumors as containing or not containing cancer cells, that is, classifying a tumor as "benign" or "malignant." This is a difficult determination in experiments that are designed to find out if a substance is carcinogenic (cancer-producing). Experiments on chemicals involve the injection or ingestion of the substance into laboratory animals, usually mice or rats. A pathologist must then determine if an animal did or did not develop cancer. There may not be agreement on what constitutes cancer in an animal so that the result of the whole experiment depends on the definition of cancer.

The smoking and health controversy presents another example of a definition and measurement problem. How can we define and measure the amount of smoke a person consumes? If we are looking at the relationship between smoking and lung cancer, the amount of smoke inhaled would be of interest. But this is an extremely difficult measurement to make. Besides, how would we take into account the relative potencies of cigarettes?

Other considerations in the smoking and health area are these: How should pipe and cigar smoking be compared to cigarette smoking? How should a person who has smoked two packs of cigarettes a day for 10 years be compared to a person who has smoked one pack of cigarettes a day for 20 years? We will return to the smoking and health controversy in chapter 7 and chapter 8.

We must consider how characteristics are defined and measured before we can properly interpret any statistical analyses based on these measurements. In fact, we must keep in mind all three of the background influences on statistics that we have considered in this chapter — the source of the statistic, the type of data upon which the statistic is based, and the definition and measurement problem.

# The Sample Survey

You will recall from chapter 1 that a sample is a part of a larger collection called the population. The sample should be selected so that it is representative of the population since information from the sample is used to make statistical inference to characteristics of the population. There are many reasons why we study only a part of a population. The most important reason is that it may be too expensive to study an entire population. Recall our discussion of this problem in chapter 1.

Let us consider the background influences that might affect our interpretation of sample survey results. The use of survey sampling is extensive. Included are preelection polls, public opinion polls, market-research surveys, television-viewing surveys, etc. Behind the reported statistic, usually a percentage, are aspects of interpretation that are mostly subjective, that is, personal or individual. The background influences on survey statistics that we will discuss are the source, the population, nonresponse, the method of contact, timing, and the phrasing of questions. Consideration of the effect of these background influences on survey results is an important beginning of proper evaluation of sample survey statistics. Without such an evaluation we will not be able to distinguish between credible survey results and meaningless survey results.

## Source of the Survey

The source of a survey is surely important, as it is for proper understanding of any statistic. There are many, varied sources of surveys. The best known are the polling organizations of Gallup, Harris, Caddell, and Yankelovich.

George Gallup maintains a position of premier pollster. Gallup received his doctorate in journalism from the University of Iowa in the early 1930's. He developed a new sampling procedure for estimating newspaper readership as part of his doctoral dissertation. The use of his new sampling technique in political polling gave birth to the modern era of the sample survey. Gallup's early success at predicting the winner of presidential elections had much to do with the public's almost unquestioning acceptance of the opinion poll. Gallup's reputation stayed remarkably intact following the polling disaster of 1948. During the 1948 presidential campaign the major polling organizations including the Gallup organization suffered a severe setback in public acceptance when Harry S. Truman beat Thomas E. Dewey contrary to what the polls had predicted.

As Gallup is generally considered to be a "Republican" pollster, Louis Harris is considered "Democratic." Harris's label came from the polling work he did for John Kennedy's successful presidential campaign of 1960. Harris's work is considered successful since Kennedy won the election. Had Richard Nixon won the presidency in 1960, his pollster, Claude Robinson, might be as well known today as is Louis Harris. Harris, like Gallup, runs a syndicated newspaper column giving results of public opinion polls.

Pat Caddell represents a third generation of political pollsters. Caddell became interested in polling through a tenth-grade mathematics-class project. At 21 he became the pollster for George McGovern's unsuccessful campaign for the presidency in 1972. He also did the polling for Jimmy Carter's successful 1976 presidential campaign.

Daniel Yankelovich gets media exposure for the survey work he does for CBS news, *Time* magazine, and the **New York Times**. The **Yankelovich Monitor**, a single copy of which sells for about $11,500, is a once-a-year publication that reports on changing American values and life styles. The Yankelovich organization also publishes **Corporate Priorities** at about $23,000 per copy. Major companies subscribe to learn about public attitudes toward the quality and the safety of products, industrial pollution, etc.

It would be futile to try and classify the many different survey organizations as good or bad. We will, instead, concentrate on judging the quality of the survey itself. Since such a judgment requires specific information, the availability of this information serves as a good criterion for judging the source of the survey. A good source will properly describe the survey results, will indicate its limitations, and will explain exactly how the data were obtained.

Such a mode of judging the source of a survey is emphasized by the American Association for Public Opinion Research, which in 1968 adopted a code of professional ethics. This code requires that a report of survey results include:

1. The size of the sample and sample design (to be discussed in chapter 5)

2. A description of the method of contact for interviews

3. A description of the population sampled

4. The timing of the interviews

5. The exact wording of questions

A word of caution about the source of a survey: Never judge survey results solely by the source. An investigation of a claim by a student newspaper that 9 out of 10 students read the paper in question led to the man in charge of advertising. His sole defense of the survey results was that the survey was conducted by a research company with a good reputation in the advertising field.

The defense of the survey results given by the man in charge of advertising was inadequate. He had not even attempted to determine if the survey results were credible. Remember: Judge the source by the survey, don't judge the survey by the source.

### *Population*

Information about the population sampled is needed to judge a survey. We must determine whether the sample is representative of the population in question. We must also be sure that inference is made to that population only. The mechanics of drawing a credible sample may have been followed so that the sample is representative of one population. However, inference may be incorrectly made to another, usually larger, population. For example, a credible study may be conducted at a university. The sample of students on which the inference is to be based may have been properly drawn and inference may correctly be made to the entire student population at the university under investigation. A serious error will be made if inference is made instead to all college students in that state or even in the country. Inference can be made only to the population sampled.

In preelection polling the population of interest does not even exist at the time the survey is run. For such surveys we are interested in the population consisting of people who vote on election day. At the time of a preelection survey, people questioned may not know for sure for whom they will vote, let alone if they will in fact vote. Inference from a sample taken before election day to this nonexistent population can be weak.

Surveys are not the only area in which we need to be cautious in this regard. Results from experimentation also need to be carefully scrutinized. As we will discuss in some detail in chapter 7, experimentation on human subjects follows animal experimentation. Experiments are conducted on laboratory animals such as mice, rats, and guinea pigs. Inference from a sample of, let's say, rats to the human population is clearly quite weak. If a chemical produces cancer in laboratory animals, it is labeled carcinogenic. Does this mean that humans will necessarily develop cancer from a carcinogenic substance?

Inference to the wrong population can lead to the statistical doublespeak we discussed in chapter 1. Doublespeak results when sample results are claimed to be representative of a population that is different from the population actually studied. In the case of surveys it is not at all uncommon for inference to be made to a population larger than the population actually sampled.

An example of this type of misuse of statistics appeared in a student newspaper that commented, "Among the 70 percent nonreplies [to a questionnaire on abortion] we can be 95 percent certain that less than 10 percent support no change [in present abortion laws]." Inference is being made here to individuals not returning a questionnaire. Quite clearly, inference cannot be made to individuals not replying.

### Response Rate

Rate of response is another factor that influences interpretation of survey statistics. Usually, a portion of a sample does not participate in a survey. Reasons for such nonparticipation include refusal to answer questions or an interviewer's inability to contact a potential respondent. Nonrespondents are representative of a part of the population for which there is no information. For example, if 30% of a sample does not respond, we have no information on the 30% of the population that the nonrespondents were to represent. Inference can be made only to the part of the population represented in the sample (70% of the whole population in this case). Inference to the entire population is not proper. Here lies one of the greatest problems in the proper interpretation of survey results: *we are often not told response rates*. But we must know them. The American Association for Public Opinion Research has not helped in this regard.

Surveys with high nonresponse rates are the so-called *voluntary response surveys* in which people exposed to a survey decide for themselves whether they want to be in the sample. The most common type of voluntary response survey is the mailback survey. In this case a questionnaire is mailed to a person or is printed in a newspaper. Readers of the questionnaire then choose to be in the sample by mailing in a completed questionnaire. An important question is, "What population do the respondents represent?" There is usually no answer to this question.

There is only one way to get credible results from a voluntary response survey. Those who do not return the questionnaire need to be contacted by telephone or in person. (Contacting a subsample of nonrespondents would suffice.) If this is not done, the survey results must be viewed with suspicion. The greater the nonresponse rate, the more suspicious we must be.

### Method of Contact

Another important factor needed to judge a survey is the method of contact used to collect the data. For example, respondents may have been contacted by telephone. This necessarily reduces the population studied to people with access to a telephone. Since many telephone surveys rely upon numbers selected from a telephone directory, the population is further reduced to people with listed telephones. This method can be especially restrictive in metropolitan areas where many telephone numbers are not listed. For example, as many as 50% of the telephones in the Washington, D.C. metropolitan area are not listed. The opinions of people in unlisted households can be quite different from the opinions of people in listed households. Such differences might vary from question to question, but they need to be considered.

As another example, consider participants in a television-viewing survey or product-research survey. They may first have been contacted by telephone and asked to keep a television log or to test a product at home. Such a technique reduces the population to households with (listed) telephones, just as in the case of an actual telephone survey.

## Timing

Another background influence on survey results is the timing of a survey. For example, Gallup reported the public's approval of Congress at a rate of 48% in August of 1974, which is a large increase over the April 1974 level of 30%. Between April and August of that year the House Judiciary Committee had voted articles of impeachment against President Nixon. This vote had a profound effect on the public's view of Congress.

So important is consideration of the timing of television-viewing and radio-listening surveys that the Federal Communications Commission proposed rules prohibiting station contests within four weeks of a rating survey and mass mailings within four weeks of a survey.[14]

Be careful when considering the effect timing might have on survey results. Too often timing is given too much influence. Invariably, when two surveys disagree, the timing of the surveys will be blamed. Don't be too anxious to let the timing of surveys be the scapegoat for improperly designed and/or conducted surveys.

## The Question

We continue our evaluation of survey results by considering another important background influence on survey results, the question. The phrasing of the question can affect survey results. Some questions are constructed more as a statement than as a question. Examples include: "Should the insurrection of long-haired college students be stopped now or later?" or, "Would you favor a no-fault insurance plan that would increase your premiums by 30%?"

© 1972 United Features Syndicate.

Survey questions may be closed or open. Closed questions are ones in which a respondent is asked to choose between one of a fixed set of responses. As Lucy indicates in the Peanuts cartoon shown here, a person's response is necessarily restricted to the response categories available. This restriction makes the "no opinion," or "don't know," category a catchall for many different responses. Other reasons for a "no opinion" response are:

1. The wording of the question confuses the respondent.

2. The respondent does not wish to commit himself or herself because of lack of confidence or lack of knowledge.

3. A knowledgeable respondent does not have an opinion on the subject.

The wording of survey questions was an important issue relative to gauging public opinion during two recent episodes in American history, the Vietnam War and the Watergate affair. Public opinion was a crucial variable in the determination of government policy during both the conduct of the war and the investigation of the Watergate scandal. However, public opinion appears to vary when questions on the same issues are phrased differently.

The Vietnam War became a battle for public opinion. Many feel that North Vietnam and South Vietnam as well as the United States tied battlefield tactics to United States public opinion. In 1967 two congressmen ran surveys in their respective districts concerning a decision by President Lyndon B. Johnson to extend bombing to the North Vietnamese cities of Hanoi and Haiphong, both key links in the northern supply network. One congressman asked, "Do you approve of the recent decision to extend bombing raids in North Vietnam aimed at the strategic supply depots around Hanoi and Haiphong?" Sixty-five percent supported the president's action when the question was so worded. The second congressman asked, "Do you believe the U.S. should bomb Hanoi and Haiphong?" This time only 14% said "yes."

It is hard to say what part of this difference in opinion is due to the use of the phrase "strategic supply depots" since these surveys were of two different congressional districts. Certainly, there must have been some effect from the use of that phrase.[15]

The Watergate episode was also closely tied to public opinion. As the House Judiciary Committee considered articles of impeachment against President Nixon, the committee members kept an eye on the public's feelings on this important issue. Here again, the wording of the questions asked in opinion surveys was critical. Gallup asked the question, "Do you think President Nixon should be impeached and compelled to leave the Presidency, or not?" Only 30% thought Nixon should be impeached when impeachment was tied to removal from office. When people were asked in a private Caddell survey, "Do you think the President should be tried and removed from office if found guilty?" 57% said yes.[16]

A large part of the difference found by these two surveys can be attributed to the wording of the questions that were asked. One reason for the strong effect that the phrasing of the questions had on response in this case is that people just did not understand the impeachment process.

## Survey Doublespeak

Recall from our discussions in chapter 1 that statistical doublespeak refers to the misuse of statistics. By survey doublespeak we mean doublespeak resulting from the improper use of credible survey data or the use of meaningless survey data.

Survey doublespeak could well be a result of one or more background influences working together. For example, in late 1975 Gallup reported that the public opposed federal aid to New York City 49% to 42%. A telephone survey ran jointly by CBS and the *New York Times* found the public in favor of aid 55% to 33%. The difference: the Gallup poll (probably conducted by personal interview) was run before a speech in which President Gerald R. Ford spoke out against aid. The CBS–*N.Y. Times* poll came after this speech. The method of contact and the timing of these two polls could explain the doublespeak.

Another example of survey doublespeak that can, for the most part, be explained by background influences was a contradiction between the Gallup and Harris polls. Harris reported in late 1975 that President Ford trailed Senator Hubert Humphrey 52% to 41% when matched in a head-to-head race for the presidency (1976 election). Gallup, on the other hand, found Ford ahead of Humphrey 51% to 39%. All else being about the same, the timing of these two surveys might explain the doublespeak. The Harris survey was taken after the president announced a cabinet shake-up and after former California governor Ronald Reagan announced he would oppose Ford for the Republican nomination but before Ford traveled to China. The Gallup poll, in contrast, came after Ford left Peking. An increase in presidential popularity often follows foreign travel, but this unusually large variation in presidential support over a short time interval indicates that the public's opinion of Ford had not yet stabilized.

So we see that timing can be an important factor in our interpretation of sample survey statistics. But, as mentioned before, timing can too often be blamed. When confronted with conflicting statistics, two pollsters will often blame the difference on the timing of the surveys when, in fact, one or both of the surveys may be improperly designed and/or conducted. We will have to evaluate the influence of timing on sample survey results, and be suspicious of claims of the effect of timing on the statistics.

Many surveys may require consideration of only the background influences we have discussed: the source, the population studied, the response rate, the method of contact, the timing, and the questions asked. Other considerations of interpretation of survey results will be discussed in chapter 5.

## Summary

Interpretation of statistics based on the considerations discussed in this chapter is for the most part subjective or personal. There is no right way to judge these influences on statistics. The important thing is that we consider them. They are important!

The source of a statistic may be a clue as to the credibility of the statistic. We must also be aware of the type of data on which the statistic is based. Both the description of a set of data and the inferential techniques we will discuss vary for different types of data. We must also be aware of how key characteristics are defined and then measured. By considering the background influences of source, definition, and measurement as well as the type of data we can do a lot to determine whether a statistic is credible.

We encounter many survey statistics in the media. We must be sure to consider all of the following background influences on these statistics:

1. The source of the survey statistics

2. The method of contact for interviews

3. A description of the population sampled

4. The exact wording of questions

5. The response rate

A consideration of these background influences will go a surprisingly long way toward distinguishing between meaningful and meaningless survey information.

Much of the information that we need to judge a statistic may not be available. (Recall our desire to know the response rates of surveys.) If this information is not available, we must, of course, be suspect of the resulting statistics. Somehow, if a statistic is reliable, this background information is usually made available.

Let us remember these potential influences on statistics as we begin to look at the numbers themselves. We'll often feel like the tax accountant who lamented as his client was sent to jail for tax evasion, "I did the best I could with the information I was given." Let us see what we can do with the statistics we encounter daily. Let us see what they might mean.

# No Comment

■ *"If we put aside the purely statistical element involved in the* Redbook *report, we can say that the findings you (*Redbook*) offer seem significant, and judging by our clinical experience confirm much that we have been observing in recent years."*
*(W. M. Masters and V. E. Johnson concerning a* Redbook *sex survey)*

■ *"There's little doubt that the Government figures are honest. The problem is that the things the numbers are supposed to measure cannot be determined with precision.*
*"Furthermore, the problem is compounded when the figures are reported to the public. Unless one is aware of the pitfalls and few people are — the statistics can become extremely misleading."*
*(T. F. O'Leary, Jr., 1975)*[17]

# Exercises

1.  What are the background influences (source, measurement, and definition problems) on FBI crime statistics based on the following description of these figures?

    The crime rate reported by the FBI on a yearly basis is the number of crimes reported to local law enforcement agencies per 100,000 population. The 7 major crimes included in this statistic are:
    a.  Forcible rape and attempted forcible rape
    b.  Aggravated assault and attempted aggravated assault
    c.  Murder — defined as excluding deaths caused by negligence, suicide, accident, or justifiable homicide
    d.  Robbery or attempted robbery — defined as taking something of value by force or threat of force in the presence of the victim and usually resulting in injury to the victim

  e. Burgulary — defined as entry of a structure to steal, even if no force is used to gain entry

  f. Larceny — defined as taking of property without force and including shoplifting, pocket-picking, purse-snatching, auto parts theft, or bicycle theft but *excluding* embezzlement, "con" games, forgery, or issuing bad checks

  g. Auto theft, attempted auto theft, and temporary use of an auto

   Local police agencies either report on a voluntary basis directly to the FBI or, as is the case with about 14 states, to a state agency that sorts the data and then sends the information to the FBI. The local crime rates are used to determine federal aid for local police agencies and are used, along with arrest figures, as an indication of the effectiveness of various law enforcement agencies.

2. A local telephone company mailed a questionnaire to all its customers along with the monthly bills. The questionnaire concerned extension of the local service area. Assuming that all of the company's customers received such a questionnaire, what would be the advantages and disadvantages of such a mailing to a credible sample survey?

3. What are the possible reasons for the following example of statistical doublespeak?

  a. 38% of Americans wanted President Nixon to remain in office (*Time*–Yankelovich telephone survey run 15 and 16 May 1974)

  b. 41% (Harris poll conducted 13 to 17 May 1974)

  c. 55% (questionnaires mailed from the White House that were returned between 29 April and 10 May 1974; of the 6,000 newspaper editors, broadcasters, and White House supporters sent the questionnaire, 1,677 responded)

4. Some respondents to a survey may consider the answers to some questions private. For example, questions on income, education, or sexual habits are sensitive. Should the results of such questions be judged differently from the results of less sensitive questions? How might valid information be obtained on such topics?

5. A local newspaper ran a *Your Opinion Counts* section in its Saturday edition. Unedited questions sent in by readers were published. Readers would clip this article and send in their opin-

ion on the questions asked. Results (just percentages, not the number responding) were published the following week with a series of new questions. How representative of public opinion in the newspaper's circulation area is such a straw vote? How should not knowing the number of respondents affect interpretation of the results?

6.    A mailback newspaper questionnaire asked readers who their two favorite television personalities were. No separate categories for local or national stars were given. After returns "were digested," these two response areas were separated. The top national and the top local television personality were then reported. How valid is such a procedure, irrespective of the credibility of the original survey?

7.    Two statewide surveys were to be run at the same time by groups at two state universities. Questions concerned public opinion on a variety of topics. One survey was a telephone survey and the other was a mailback questionnaire. What procedures might ensure valid results in each case? Would the results be comparable, let's say, for the state in which you live?

8.    A professional polling organization found that 43% of the residents of a state favored a judicial reform amendment. Nine percent were opposed and the rest undecided. Will the amendment become part of this state's constitution?

9.    In rebuttal to arguments put forth by various consumers and parents groups, the Association of National Advertisers commissioned a survey. Mothers were asked if it was a benefit to children to acquaint them with products through commercials. Forty-one percent felt it was beneficial, and 43% felt it was not. What might have influenced these statistics besides the selection of a representative sample?

10.   A "poll" was conducted in 7 of 12 dormitories at a university concerning student opinion on the student government's recognition of a gay students coalition. Questionnaires were passed out to the 2,190 students living in these 7 dorms, with 1,327 responding. Of those responding 1,015 "opposed," 175 "approved," and 137 had "no opinion" on the student government's action.

   a.    Discuss the claim that the sample is quite representative of the student body.

b.  Discuss inference to the following populations: all students living in the 7 dorms; all students living in the 12 dorms; and the entire student body at this particular university. Are the results of this survey representative of any group(s) on campus?

# For Discussion

1.  Discuss the definition of a SMSA. Consider examples near you. Is it surprising that a large proportion of land in SMSAs is rural and that a large proportion of citizens live in SMSAs?

2.  Discuss background influences on the following television-viewing surveys:

    a.  Arbitron:

    The sample is selected from telephone directories except in areas with a high concentration of black or Spanish-speaking people. Households are asked to keep a record of which channel is being watched and who are watching at 15 minute intervals.

    b.  Neilsen:

    Samples are supposed to include both telephone and nontelephone households.

    *Sample #1:* Includes a fixed number of about 1,200 households with an audimeter on each television set. The audimeter automatically records tuning information, which is stored and automatically transmitted by phone lines to a central computer.

    *Sample #2:* Records (audilogs) are kept manually as for Arbitron. A recordimeter is used. This device is installed on television sets. It records hours of use and provides reminder signals for prompt entry of information in the audilog.

3.  In June 1970 *Consumer Reports* ranked auto insurers on the basis of 230,871 returns from the more than 1.4 million CU members who were mailed the 1969 annual questionnaire. Less than half of the five-page (77 questions) questionnaire concerned CU members as auto insurance policyholders and liability claimants. One of America's largest auto insurers was judged "distinctly inferior" to the other 25 companies that were evaluated. Discuss how this auto insurer might refute CU's ratings. Is a rebuttal possible by CU? Can these 25 auto insurers be reasonably ranked by such a process?

4.  Ask the members of your class certain thought-provoking open questions. Then, have at least two different members of the class categorize the answers. Any variation in the classification of responses?

5.  Discuss the definition and measurement problem in the smoking and health area, which we briefly mentioned in the section on definition and measurement considerations in this chapter.

6.  Relative to preelection polling in primary elections discuss inference from the sample to the nonexistent population of those who will vote on election day. How difficult would it be to try to determine if a person is going to vote?

7.  Ask your classmates a closed question on a thought-provoking topic. Are answers easy to get? How does this experience affect your interpretation of the results of a survey that seeks the answers to closed questions? Why do you think the news media might prefer to write about surveys that asked closed questions? (Tie your answers to Discussion 4.)

8.  Discuss the anonymity of respondents in a mailback survey with respect to the need to recontact nonrespondents.

9.  Discuss why response rates for telephone surveys might be quite low.

10. The unemployment rate for both January and February, 1975 was 8.2%. During this period 580,000 persons dropped out of the labor force. This presumably occurred because people became discouraged and quit looking for work. Including these 580,000 individuals in the labor force would have made the February rate 8.8%.

     From June to July 1976 the unemployment rate jumped from 7.5% to 7.8%. Good news? During this period 700,000 people *entered* the labor force with 400,000 finding jobs.

     Discuss the definition of unemployment and employment with regard to these two examples. Why might people have entered the labor force in July of 1976?

# Further Readings

There are a few books that address themselves to the background influences on survey statistics that we have discussed. You'll enjoy a look at politics and the sample survey in both of the following: *Polls: Their Use and Misuse in*

*Politics* by C. W. Roll, Jr. and A. H. Cantril or *Lies, Damn Lies, and Statistics* by M. Wheeler. Both publications are cited in the Notes to this chapter.

Books about the misuse of statistics are also likely to supplement our discussions because the misuse of statistics is often due to background influences. Books that discuss the misuse of statistics are: S. K. Campbell's *Flaws and Fallacies in Statistical Thinking* (Englewood Cliffs, N.J.: Prentice-Hall, 1974), D. Huff and I. Geis's *How to Lie with Statistics* (New York: Norton, 1954), and R. S. Reichard's *The Figure Finaglers* (New York: McGraw-Hill, 1974).

# Notes

1. C. W. Roll, Jr., and A. H. Cantril, *Polls: Their Use and Misuse in Politics* (New York: Basic, 1972), chapter 1.
2. *Ibid.,* p. 13.
3. J. W. Duncan, Chairman, ASA-FSUC Committee on the Integrity of Federal Statistics, "Maintaining the Professional Integrity of Federal Statistics — A Report of the ASA-FSUC Committee on the Integrity of Federal Statistics," *American Statistician* 27 (1973): 58.
4. *Facts on File Yearbook* 31 (New York: Facts on File, Inc., 1971): 212.
5. P. M. Hauser, "Statistics and Politics," *American Statistician* 27 (1973): 69.
6. R. Nader, *Beware* (New York: Law-Arts, 1971), pp. 66–67.
7. W. C. Magnuson and J. Casper, *On the Dark Side of the Marketplace: The Plight of the American Consumer* (Englewood Cliffs, N.J.: Prentice-Hall, 1968), p. 128.
8. W. A. Wallis and H. V. Roberts, *Statistics: A New Approach.* Copyright 1956 by the Free Press, A Corporation (New York: Macmillan, 1956), p. 69.
9. R. L. Stein, "New Definitions for Employment and Unemployment," *Employment and Earnings and Monthly Report on the Labor Force* (February 1967): 3–13.
10. U.S. Department of Labor, Bureau of Labor Statistics, "Concepts and Methods used in Manpower Statistics from the Current Population Survey," Report 312 (1967): 4.
11. L. H. Cronbach, *Essentials of Psychological Testing* (New York: Harper and Row, 1949), p. 107.
12. H. H. Kendler, *Basic Psychology* (New York: Appleton-Century Crafts, 1963), p. 607.
13. W. W. Willingham, "Predicting Success in Graduate Education," *Science,* vol. 183, no. 4122 (1974): 274–275.
14. D. Reed, "New Rules Concerning Ratings, Contests," Lexington (Ky.) *Lexington-Herald-Leader* "T.V. Spotlight" (6 July 1975), p. 8.
15. M. Wheeler, *Lies, Damn Lies, and Statistics* (New York: Liveright, 1976), p. 138.
16. *Ibid.* p. 167.
17. T. F. O'Leary, Jr., "Government Statistics — Can You Believe All That You Read?" *U.S. News and World Reports* (10 March 1975), pp. 64–65.

# Natural and Descriptive Statistics

3

IN THIS CHAPTER we will discuss both natural (census) and descriptive statistics. In 1790 the United States became the first country to take a periodic census. Many different censuses are presently conducted by the Bureau of the Census. Much additional information is obtained by the bureau through the sampling of many different populations. Only a fraction of data collected by the Census Bureau results from attempts to contact an entire population.

The descriptive statistics considered in this chapter are graphs, indicators of the middle of a set of data (called indexes of central tendency), and indexes of the spread of data about the middle (called indexes of dispersion). The graphic techniques that are likely to be used by the media are discussed. We consider how graphs can be used to properly describe how data are arranged (the distribution of data). We will also see how graphs can give a false picture of the information contained in a collection of numbers.

Knowing the distribution of a set of data will allow us to properly describe the middle of the data as well as the dispersion of data about the middle. Learning to properly describe a set of data will enable us to be able to detect any misuse of descriptive statistics.

As we proceed with our discussions, do not forget that consideration of the background influences that we discussed in chapter 2 may lead us to feel that certain numbers are meaningless. Assuming that we determine a number worthy of further consideration, in this chapter we will see how to properly evaluate a natural or a descriptive statistic.

## Natural Statistics: The Census

As mentioned in our discussion of the history of statistics in chapter 1, the first statistics were natural, or census, statistics.

> *Natural*, or *census, statistics* are numbers that are collected through an investigation of an entire population.

43

The earliest documentation of census taking in the Western world was in Babylonia around 2800 BC, but census taking may go back as far as 4500 BC. Enumerations in China date back as far as 3000 BC. There are many Biblical references to census taking. Most notable is a 1017 BC census by David, the Hebrew king. There are indications that the Biblical description of the Divine wrath following David's enumeration of the Israelites gave rise to the idea that census taking was a religious offense. This may have delayed adoption of the census of England for many years.[1]

In 1787 the United States Constitution provided for the first periodic enumeration of a population. "The French statistician, Moreau de Jonnes, declared that the United States presents a phenomenon without parallel in history — 'that of a people who instituted the statistics of their country on the very day when they founded their government.' "[2] Section 2 of Article 1 of the Constitution requires that:

> Representatives and direct Taxes shall be apportioned among the several states which may be included within the Union, according to their respective Numbers. . . . The actual Enumeration shall be made within three years after the first meeting of the Congress of the United States, and within every subsequent Term of ten years, in such Manner as they shall be Law direct.

The first enumeration of the nation's populace took place on the first Monday in August 1790, less than a year after the inauguration of President Washington. The first censuses were quite crude by modern standards. Marshalls of the United States judicial districts used limited funds to collect data on 6 items: the name of the head of each household; the number of persons in each household who were free white males of 16 years and upward; the number of free white males under 16 years; the number of free white females; the number of other free persons; and, the number of slaves in each household. Some of the more important developments of the United States Census from 1790 to 1970 are summarized in table 1.

The mail service was extensively used for the collection of data for the 1970 census. Approximately 60% of the population (people living in metropolitan areas) received questionnaires through the mail and were asked to mail them back to the nearest district census office. Enumerators visited households from which questionnaires were not returned or were improperly completed. For the other 40% of the population (people in rural areas or small towns) census questionnaires were left by the mailcarrier. An enumerator then visited each household to assist in the proper completion of the questionnaire.

Only the 7 questions mentioned in table 1 for the 1960 census were asked of everyone in the 1970 Census of Population: name, address, age, sex, color or race, marital status, and relationship to head of

**TABLE 1** Historical Development of the U.S. Census 1790–1970

| Census | Changes in Data Collected and Censuses Conducted | Innovations |
|---|---|---|
| 1790 | Census of Population | |
| 1810 | Census of Manufacturing | |
| 1820 | Inquiries into commerce and agriculture | |
| 1830 | Inquiries into school attendence, illiteracy, and occupation: first Census of Agriculture (every 9 years today) and Census of Mineral Industries | Uniform printed schedules |
| 1840 | | Household not individual became unit of enumeration |
| 1850 | Inquiries into taxes, schools, crime, wages, value of estates and mortality statistics: first Census of Governments (local, state, etc.) | Results reported for "known civil divisions" as counties, townships, etc. |
| 1870 | | Rudimentary tabulating machine used |
| 1880 | Privacy of information first protected: first Census of Religious Bodies | Use of enumerators and supervisors working for Superintendent of the Census rather than using marshalls and their assistants |
| 1890 | Farms and home mortgage data collected as well as information on corporate and individual indebtedness | Separate schedule for each family, use of punch cards and electric tabulating machines |
| 1910 | | Permanent Office of the Census with competitive testing of enumerators |
| 1930 | Census of Unemployment and Census of Business | |
| 1940 | Income, internal migration: first Census of Housing | Use of sampling techniques |
| 1950 | Survey of Residential Financing | Use of electric computer |
| 1960 | Census of Transportation | Extensive sampling (Only population questions concerning address, age, sex, color or race, marital status, and relationship to head of household are asked everyone.) and use of the mails |
| 1970 | Occupation 5 years before census | Mails extensively used with responses "read" electronically |

Source: Adapted from *Working Paper 39*, U.S. Bureau of the Census, and *Fact Finder for the Nation*, U.S. Bureau of the Census.

household. Other questions were asked of samples of 5% and/or 15% of the households — that is, some questions were asked of a 5% or a 15% sample, and some questions were asked of both samples (a 20% sample). Only 15 housing questions were asked of each household for the Census of Housing, which is conducted simultaneously with the Census of Population.

In 1970 data summaries were published for different geographic areas including approximately 750,000 blocks in certain large metropolitan areas. Summary statistics are available for areas no smaller than a block, that is, individual household data is kept confidential.

As indicated in table 1, the United States does not run one census but many censuses. Survey information is collected on all the topics for which the government needs information to make policy decisions. The census we are most familiar with is the decennial (every ten years) Census of Population and Housing. But surveys are run by the Census Bureau between the decennial Census of Population and Housing as well as at the time of this periodic attempt at complete enumeration.[3]

The United States Census is not a census in the strictest sense since not everyone is enumerated. The Census is, however, as accurate as possible for the money spent. The point of diminishing returns has been reached — that is, the cost of increased accuracy is much greater than the amount of information that would be obtained. The *undercoverage,* or people not contacted by a 100% sample, has been estimated to be 2.5%.[4] Undercoverage is, as one would expect, not uniformly spread over the nation's populace, as shown in table 2.

Census data form the foundation for much of the statistical work done in the United States. Government statistics and survey designs are very closely tied to the information that is obtainable from the Census Bureau. For example, geographic boundaries used in the design of a sample survey often agree with census boundaries. Included in the census are the summaries for states, counties, census tracts, enumeration districts, and blocks. For such areas census information necessary

**TABLE 2**    **Undercoverage of 1970 Census of Population**

| Sex | Omission Rate (%) | | |
| --- | --- | --- | --- |
| | All | White | Black |
| Both sexes | 2.5 | 1.9 | 7.7 |
| Male | 3.3 | 2.4 | 9.9 |
| Female | 1.8 | 1.4 | 5.5 |

Source: U.S. Bureau of the Census, Series P-23, No. 56, 1975.

for the design of a credible sample survey is available. This information includes the size of these areas. As we will discuss in chapter 5 and chapter 8, proper representation of a particular geographic region in a sample might require that the number of sampling units taken from the region be proportional to the size of the region. Only from census records can these sizes be determined.

Legislation that calls for a census of the United States population every 5 years rather than every 10 years has long been considered. The 1980 census may be followed by a 1985 census. Since the Census of Population and Housing provides basic information for the determination of many government and survey statistics, up-to-date census information will lead to generally more accurate statistics.

## Graphic Displays of Data

In 1786 at the age of 27 William Playfair published a volume called **Commercial and Political Atlas** in which he represented "by Means of Strained Copper-Plate Charts the Progress of the Commerce, Revenue, Expenditures and Debts of England." Concerning this first use of graphic statistics Playfair said, "I have succeeded in proposing a new and useful mode of stating accounts. . . . As much information may be obtained in five minutes as would require whole days to imprint on the memory, in a lasting manner, by a table of figures."[5]

". . . As you can see, the profit picture for oil companies isn't THAT bright . . ."

> ***Graphic statistics*** (for example, graphs, charts, and pictures) are devices for representing or summarizing numerical information.

Data that have been collected are commonly presented through the use of graphs, charts, or diagrams. Since the census of 1870, census data have been summarized by using such devices. As we consider graphic statistics, do not forget that proper interpretation of information conveyed by graphic techniques requires a consideration of the background influences discussed in chapter 2.

### Histograms

As we begin to learn how to understand the information contained in graphs, let us first formulate what we mean by terms that are used to describe the form of a distribution of numbers in a data set. The graphs in figure 1 are ***histograms***, a method of graphing both

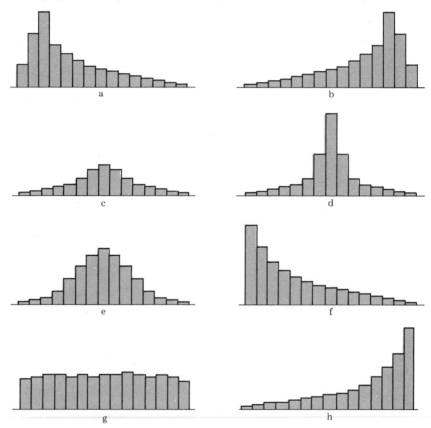

FIGURE 1    The Form of Distributions of Numbers

discrete and continuous data. (The term *bar graph* is used to describe a similar graph of categorical data. In a bar graph the bars are separated, not connected as in the histogram. The separation of the bars of the bar graph reflect the separate categories of such data.)

The histogram gives a pictorial representation of the proportion of data that occurs in various areas of the data scale. The horizontal axis of the histogram indicates data values. The height of the bars of a histogram represents either frequency (number of data) or the proportion of occurrences of the corresponding data values. If the bases of the rectangles in a histogram are not equal, adjustments in the vertical scale must be made. The *areas* of the rectangles must represent the proportion of occurrences of the corresponding data values (see, for example, figure 2).

## *Distribution*

The distribution of the data graphed in figure 1(a) is described as being *skewed to the right*, or *positively skewed*. Figure 1(b) is *skewed to the left*, or *negatively skewed*. Figures 1(c), 1(d), and 1(e) represent data with *bell-shaped symmetric* distributions with varying degrees of flatness or peakedness. The word *symmetry* refers to distributions whose upper and lower halves are mirror images of each other. The data graphed in figure 1(f) is *L-shaped*, while figure 1(h) represents a *J-shaped* distribution. Distribution *g* in figure 1 is referred to as being *rectangular*, or *uniform*. Distributions *a, b, c, d, e, f,* and *h* are said to be *unimodal* as there is a single peak or mode.

Graphs can be misleading or used to mislead, as indicated by the political cartoon at the beginning of this section. But a graph can provide an informative picture (description) of a set of data. To illustrate the usefulness of a graph, consider the graphs in figure 2. In contrast to the histograms in figure 1, the bases of the rectangles in the histograms in figure 2 vary in size. The vertical scale is adjusted so that the areas of the rectangles equal the percentage of people in the corresponding portions of the data scale. Recall that the defining characteristic of histograms is just that: The areas of rectangles must be proportional to the frequencies of the corresponding data values.

We could describe educational achievement as measured by years of school completed by United States citizens as *bimodal* — that is, two peaks, or modes, occur in the graph. There is a major, or most pronounced, mode at 12, which represents a high school education. A minor, or less pronounced, mode exists at 8, which represents a grade school education.

Both household size and age of the population are shown in figure 2 as skewed to the right (see figure 1[a]).

Household size is discrete (ratio) data, and educational achievement and age are continuous (ratio) sets of data. We record age, for

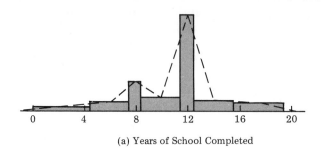

(a) Years of School Completed

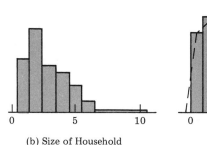

(b) Size of Household

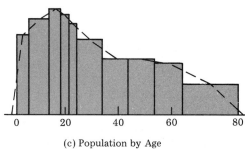

(c) Population by Age

**FIGURE 2** **Graphs of Data**

Source: 1975 *Statistical Abstract of the U.S.*, U.S. Department of Commerce, Bureau of the Census.

example, to the last birthday or discretely (0, 1, 2, etc.), but the exact ages of two people can vary by the smallest amount of time. Educational achievement is crudely measured by years of school completed although education is a continuous process.

### *Frequency Polygon*

Continuous data sets are often graphed by means of a frequency polygon. Like the histogram, the frequency polygon is a graphic display of the proportion of data in varying areas of the data scale. The dotted lines of figures 2(a) and 2(c) are the frequency polygons for the continuous data sets graphed. With the polygon we see gradual changes in frequency (the vertical scale) with gradual or continuous changes in the horizontal data scale. In contrast, the histogram shows jumps at certain units. Such jumps represent changes in discrete data since discrete observations do change in "jumps."

## Semi-log Graphs

The graphs just discussed are *arithmetic graphs*. In arithmetic graphs equal vertical distance represents equal changes in the quantity graphed. Sometimes, however, it is legitimate to adjust the vertical scale of a graph. An example of a legitimate adjustment is *semi-log scaling* used for graphing the rate of increase (or decrease) of indexes such as the Consumer Price Index (CPI). A semi-log graph is shown in figure 3.

Observe in figure 3 that the distance that represents a change from 100 to 105 is *not* the same distance that represents the 5% difference between 130 and 135. In semi-log graphs equal *rates* of change are represented by equal (vertical) distances on the graph. The percentage of increase, or *rate of increase,* is found by adjusting the difference between levels of an index to the earlier level — that is, the increase over two levels of the index is the difference between the levels divided by the original level. This number is usually presented as a percentage — the rate of increase. For example, a change from 130 to 135 is about a 4% rate of increase.

$$\frac{135 - 130}{130} = .04 \text{ or } 4\%$$

In our example if the CPI were 130 at the beginning of a year and 135 at the end of the year, the CPI increased 4% for that year. A change from 100 to 105 is a 5% rate of increase. Hence, in a semi-log graph the distance from 130 to 135 on the vertical scale is 4 units as compared to 5 units that indicate the rate of change from 100 to 105.

When comparing expenditures between groups or individuals, semi-log graphing can be informative. This is especially true when the groups vary in their total assets. For example, suppose we were to use an arithmetic grid to compare changes in expenditures between a millionaire and a poor person. The changes in the poor person's spending habits would seem inconsequential. This follows because the scaling

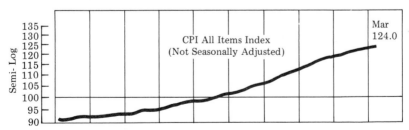

**FIGURE 3**   The Rate of Change of the CPI: 1963 to 1972
(1967 = 100)

Source: U.S. Department of Labor, Bureau of Labor Statistics, June 1972.

used would have to incorporate the large expenditures of the million-aire. Semi-log graphs would be an aid here as both people would be compared on the basis of the rate of change in their expenditures over the time period of interest.[6]

## Misuse of Graphs

Although graphs can describe the distribution of numbers in a data set, they can also give false impressions. Consider consumer prices during the period from 1958 to 1968. Figure 4 shows the changes in Consumer Price Index and Wholesale Price Index (WPI) during the ten-year period under consideration. The appearance of the change in the CPI is that of a sharp substantial increase. The same data graphed in figure 5 represent the changes in the CPI as being gradual.

The difference in these two graphs is the vertical distance used to represent 5%. A 5% difference in figure 5 is represented by a dis-

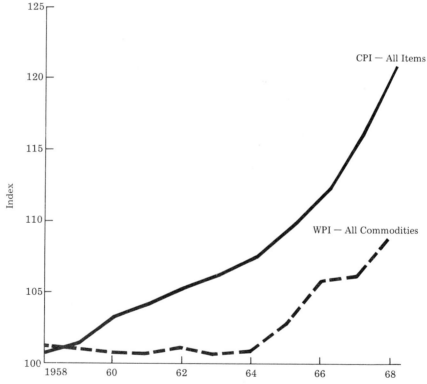

FIGURE 4    Consumer and Wholesale Price Indexes: 1958 to 1968 (1957–59 = 100)

Source: U.S. Department of Labor, Bureau of Labor Statistics, Bulletin 1647.

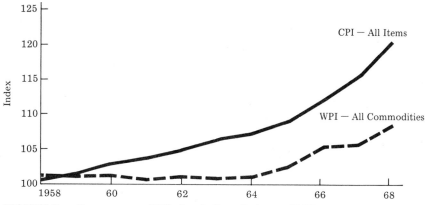

**FIGURE 5**   Consumer and Wholesale Price Indexes: 1958 to 1968 (1957–59 = 100)

tance that is one half the distance used to represent the same 5% change in figure 4. Hence, adjustments in the scales, data, and frequency or combinations of these can yield quite different pictures of the same data.

Note that the graphs in figure 4 and figure 5 have a vertical scale that does not include zero — in fact, the vertical scale for each of these graphs begins at 100. The CPI represents a change from 100, which is the base level. (We will discuss such indexes in chapter 10.) Be careful in reading graphs. If the vertical scale does not show the base value, usually zero, the bottom of the graph may have been cut off. We may therefore be looking at only the "tip of an iceberg" as much important information is not presented.

We have seen how graphs can be used to describe and to mislead. Pictorial representation of statistics can also be misleading. The comparison of oil prices in Texas, Saudi Arabia, and Venezuela, as presented in figure 6, is a case in point. The equal size of the oil barrels gives an appearance of equal oil prices, and placing the bottom of each barrel on the same line gives the appearance that the prices started at the same level in 1971. The fact that Texas prices were highest during 1971, 1972, and 1973 is masked. A bar graph of these data, as given in figure 7, permits a comparison of prices within each of the 4 years as well as a comparison of price differences between years.

Graphic presentation of data can be informative, but as we have seen, it may also be misleading. Proper interpretation of statistics that are represented by graphs requires a critical and probing eye relative to the information being presented, not just a quick glance that may lead to misinterpretation of the information contained in the graph. Other examples will help to illustrate this point.

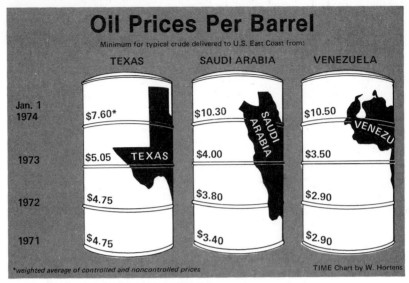

**FIGURE 6** **Pictorial Graph of Oil Prices**

Source: Reprinted by permission from TIME, The Weekly Newsmagazine;
Copyright Time Inc., 1974.

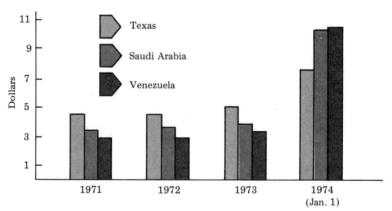

**FIGURE 7** **Bar Graph of Oil Prices**

The numbers in figure 8 represent dollars spent for home-entertainment-equipment advertisements in three of the nation's major newsmagazines. These numbers are in the ratio of 8:4:1. The radios drawn to represent these dollar amounts have their bases *and* heights in the ratio 8:4:1 — that is, the base and height of the *Newsweek* radio are 4 times larger than the base and height of the *U.S. News* radio. This makes the *areas* of the two radios different by a factor of 16! (Area

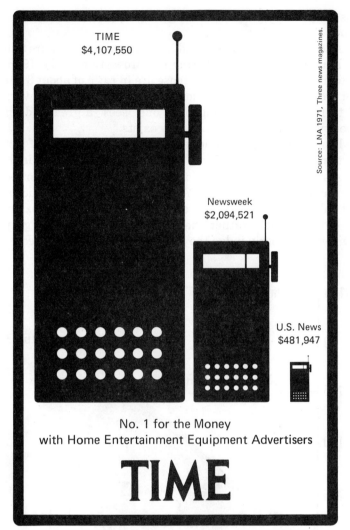

**FIGURE 8    A Newsmagazine Advertisement**

Source: Reprinted by permission from TIME, The Weekly Newsmagazine;
Copyright Time Inc., 1972.

equals base times height.) Likewise, the *Time* radio has an area 64 times greater than the *U.S. News* radio, but the dollar amounts differ by a factor of 8. A quick glance at this graph will give the impression that the numbers represented by the radios are in the same ratio as the areas of the radios. The radios have areas in the ratio of 64:16:1, giving the false impression that the amount of this type of advertising in the three magazines (*Time, Newsweek*, and *U.S. News*) is also in this ratio.

This is very misleading. Alas, such a graphing technique is all too common.

Another example was found in a newspaper. In figure 9 the areas of the human figures representing the proportion of the 1976 budget going to human resources and defense are in ratio of about 3.7 to 1. The budget projections of $152.7 billion and $94 billion are, however, in ratio of 1.6 to 1. This is misleading since one "sees" a greater discrepancy between these numbers than actually exists.

Proper graphing technique requires that the areas of pictures be proportional to the numbers they represent. This is the case in figure 10. The areas of the parts of the word "travel" in figure 10 are proportional to the numbers they represent. You will not be misled by such a graph, even if only a quick glance is given it. In any case, careful examination of the data graphed, rather than a quick glance at the graph, will enable you to properly judge the statistics being represented. (As with

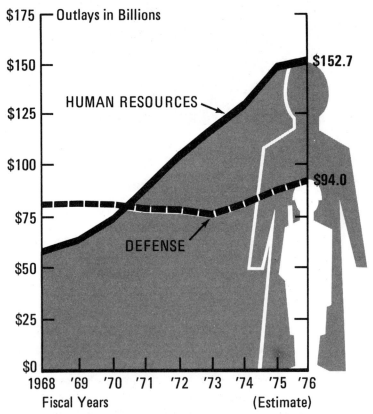

FIGURE 9    **Budget Trends: Human Resources and Defense**

Source: Copyright © 1975, *The Courier-Journal* and *The Courier-Journal & Times*, Reprinted with permission.

TIME $4,748,214

Newsweek $2,736,086

U.S. News $1,016,564

No. 1 for the Money
with Travel, Hotel & Resort Advertisers

TIME

Source: LNA 1st 6 mos. Three news magazines.

FIGURE 10    Proper Graphing Techniques

Source: Reprinted by permission from TIME, The Weekly Newsmagazine;
Copyright Time Inc., 1972.

any statistic, proper interpretation cannot ignore background influences. What advertisements are classified as "travel"?)

We conclude our discussion of the misuse of graphs by looking at
the graphing of time. Either or both of two errors might occur. First,
the representation of a year may vary. For example, in figure 11 a

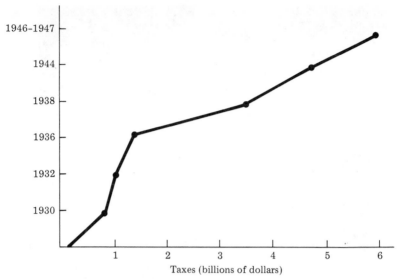

**FIGURE 11    Graphing of Time**

fixed interval on the time scale represents both 2 years and 4 years. The time period from 1946 to 1947 is represented by a point!

But a more subtle graphing error exists in figure 11. We read a graph from left to right, as you are reading this book. As we read the graph of figure 11 from left to right, we read time changes as tax changes. But taxes vary with time, not vice versa. Whenever time is graphed, it should be represented on the horizontal scale. Then we will read changes in the measurement of interest as time changes, or as a function of time. Figure 11 implies that time will change as property taxes change when, in truth, property taxes change with time.

# Indexes of Central Tendency

Our present preoccupation with averages — for example, the "average" person, the "average" family, the "average" this, and the "average" that — is in no small part due to the work of Adolphe Quetelet. In the early nineteenth century, Quetelet worked with the average man or *homme moyen*. He developed the idea that the characteristics of the average man can be represented by the mean (one type of "average") and the upper and lower limits of variation from the mean. (We will discuss variation in the next section.) Quetelet described not only the physical characteristics of man such as height , weight, and chest circumference but also the qualities of man that are not physical such as intelligence.

We will talk of three averages, or indexes of central tendency: the mean, the median, and the mode(s). It is important that you develop a sense of what these indexes say so that you may properly interpret their use and misuse.

---

*Indexes of central tendency* (averages) are statistics that describe the clustering of data.

---

By "clustering" tendency we mean the tendency of subgroup(s) of data to have the same value. If we are talking about the measurement of a human trait, we would often like to know if people have a tendency to be similar according to that trait — that is, if the numbers measuring the trait tend to be the same value.

## Median

Fechner (1874) talked of the "middlemost ordinate," but it was Galton who coined the word "median" in 1883.

---

The *median* is a number that divides a data set into two parts of about equal size; data in one part are larger in value than the median, and data in the other part are smaller in value than the median.

---

For example, the median of the discrete data set 4, 4, 5, 6, and 8 is 5, and the median of 3, 5, 7, and 12 is any number between 5 and 7, usually taken to be 6.

The previous examples involved finding the median of *arrayed data*, that is, data listed in order from smallest to largest. Consider the data set 8, 12, 4, 14, and 3. The median is 8 in this case since there are two data above and two data below 8. Hence, the median is the middle of an arrayed data set but is not necessarily the middle term in a data set that is not arrayed.

The determination of the median does not take into direct account the actual data values. We obtain the median by counting the number of data above and below a certain value, provided the data are arrayed in order from smallest to largest. A particular value is termed the median if approximately the same number of data are above and below it.

## Mean

The measure of central tendency first used must have been the mean, which was known to Pythagoras in the sixth century BC.

---

The **mean** is the sum of all the data divided by the number of data.

$$\text{mean} = \frac{\text{sum of all data}}{\text{number of items in set of data}}$$

---

The mean of the data set 4, 5, 7, 9, and 10 is 7.

$$\frac{4 + 5 + 7 + 9 + 10}{5} = 7$$

The mean, unlike the median, takes into account the magnitude of each datum. As an illustration of this, suppose that a one-gram weight is placed on a ruled stick at each of the points 1, 6, 8, 9, and 11. The weighted stick would balance at the mean (equal to 7 in this case), as shown in figure 12. The mean is the center of gravity for a set of numbers; the median is the center in the sense that the number of data above and below the median is the same.

## Mode

The word "mode" was first used by Karl Pearson in 1894.

---

The **mode** is a datum that occurs more often than surrounding data.

---

There may be more than one mode, as illustrated in figure 13, which we saw earlier in this chapter as part of figure 2. In the graph of figure 13 we see two modes. The bimodal data set in figure 13 has a major, most pro-

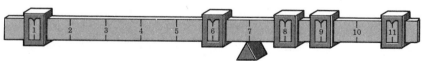

FIGURE 12    The Mean as a Balancing Point or Center of Gravity

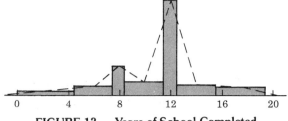

FIGURE 13    Years of School Completed

nounced, mode at 12 and a minor, less pronounced, mode at 8. (Recall that 12 represents a high school education and 8 represents a grade school education.) Since more than one mode may exist, we define a mode as a number having frequency higher than the frequencies of surrounding data. The determination of the mode can at times be subjective since what represents a "higher frequency" may be an individual or personal determination.

Observe that when more than one mode exists there is no unique "central" tendency but a clustering of data at different measurements. We would not, therefore, look for a single index of central tendency; rather, we would want to indicate all areas of clustering; that is, we would want to indicate the modes.

The mode is used to describe the clustering tendencies of categorical data. When we hear that the average American is Caucasian and female, reference is being made to the fact that more Americans are categorized as "caucasian" and "female" than are classified in other race or sex categories — that is, the *modal* race category is "caucasian" and the *modal* sex category is "female."

It is helpful at this point to observe that the word "average" may be used to refer to a mean, a median, a mode, or even another figure. Most often the word "average" refers to a mean, but, as we have just seen, an average could be a mode.

## The Relationship Between the Mean, Median, and Mode

As descriptive statistics, indexes of central tendency are used to describe the average or typical or representative value(s) in a data set. As we begin to learn how to interpret these statistics, let us look at how these numbers might compare with each other in certain instances. Referring again to figure 1, which is reproduced as figure 14 for your convenience, we will see how these indexes of central tendency might compare to each other in the eight situations illustrated.

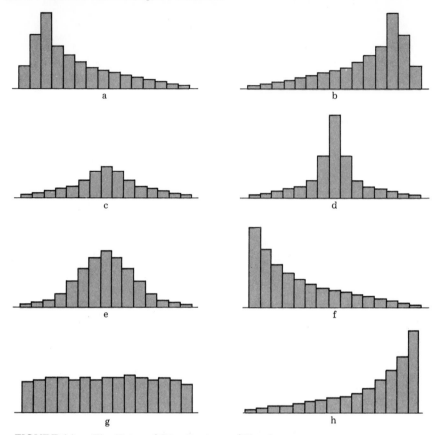

**FIGURE 14**   The Form of Distributions of Numbers

Figure 14(a) (skewed to the right) is a graph of a set of data with a few large values and a cluster or grouping of small data values. The mean, which depends on all the data, is affected by the extremely high scores and is pulled to the right of the median and mode. Hence, the mean will be the largest in value. (We can visualize this fact by recalling that the mean is the center of gravity for a data set. In figure 12 we see that if the largest datum is increased in value, the mean will be pulled to the right as the ruled stick must balance at the mean.) The mode is at the highest peak of distribution *a* of figure 14, and the median is located between the mean and the mode. Consequently, in this distribution the three indexes of central tendency are ordered (smallest to largest) as follows: the mode, then the median, and then the mean.

For example, as seen in figure 15, the data 12, 6, 4, 5, 4, 7, and 4 are skewed to the right. The mode of these data is 4, the median is 5, and the mean is 6. This ordering of the indexes (mode less than median less than mean) occurs generally in the case of data that are skewed to the right, as in figure 2(b) and figure 2(c).

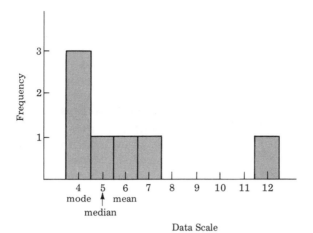

FIGURE 15    Graph of the Data (12, 6, 4, 5, 4, 7, and 4)

Distribution *b* of figure 14 (skewed to the left) is the reverse of distribution *a*. Therefore, the mean is the smallest index of central tendency, followed in size by the median, and then the mode. For distributions *c*, *d*, and *e* of figure 14 the three indexes coincide. Distribution *f* of figure 14 is an exaggeration of the situation in distribution *a*. For the scores graphically represented by distribution *g* of figure 14 a uniquely defined mode does not exist, but the mean and median will be equal and at the middle of the data scale. Distribution *h* of figure 14 is an exaggeration of distribution *b*.

We see that indexes of central tendency describe different aspects of a data set. There are times, then, when one index is a better indication of central tendency than the others. We will investigate when these indexes are properly used so that we will know when any of these statistics are misused by the news media.

### Applications

The mean is probably the most widely reported index of central tendency. However, this may be changing. For example, in a national news broadcast a newscaster was commenting on how important he felt it was that eighteen-year-olds be given the vote. (The subject matter gives you some idea of the time of this telecast.) The newscaster was trying to point out that a typical citizen of our country is young. He stated, "The median age in this country is 28 years." (In 1974 this value was 28.7 years.)

Since the distribution of ages is skewed to the right (distribution *a* of figure 14), the mean would be disproportionately affected by the ages of the elderly and would therefore not be a good representation of the typical age of a United States citizen. The newscaster may have been motivated by the fact that the best index of central tendency in this case is the median or by the fact that the median is smaller in value than the mean. The smaller number would, of course, better stress his point. (The mode would be even smaller and he would make his point even more effectively if viewers knew what a mode is.)

How, then, should we select an index of central tendency? By knowing the proper selection procedure we can better interpret statements that employ one or more of these three indexes of central tendency. Roughly speaking, if we are at all interested in the *total* of the scores, the *mean* is the best description of central tendency because the mean, as we have seen, takes into account each datum value and hence the total of all the data. (The mean is found by dividing the total of the data by the number of data.) If, on the other hand, we are interested in an index that indicates the most *typical* or *representative* data value, any one of the indexes would be the proper indication of central tendency. The choice in the latter instance depends on the distribution of the data.

In cases in which the distribution of the data is unimodal and symmetric or nearly so, as in graphs *c*, *d*, and *e* of figure 14, the mean is usually used to describe central tendency. Observe that for data that are nearly symmetric and unimodal the mean, median, and mode are nearly equal. We cannot, therefore, choose a "wrong" index of central tendency.

The median is the best index of central tendency in cases in which the data are unimodal and skewed and we are not interested in the data total. To illustrate this point, consider the numbers 6, 4, 5, 6, and 4. The mean and median are 5. If one 6 is changed to a 31, giving us the numbers 4, 4, 5, 6, and 31, the median remains 5 but the mean becomes 10. The mean is not representative of the central tendency of the numbers 4, 4, 5, 6, and 31, as can be seen by the graphs in figure 16.

Finally, for categorical data and multimodal (many modes) data the mode(s) best expresses the general location or clustering of the data — that is, the mode(s) is the best indication of central tendency.

In the example concerning the average age of a citizen, the median would best indicate the age of a typical citizen. (Recall that ages are a unimodal skewed set of data.) If we are interested in comparing the incomes of two areas as they relate to, let's say, the tax base of the areas, the mean is most descriptive since the total income of an area determines its tax base. In contrast, if we are interested in comparing the incomes of two areas in an attempt to compare the living standards of the areas, the distribution of the incomes would determine which index is most descriptive: the mean for unimodal symmetric data; the median for unimodal skewed data; and the modes for

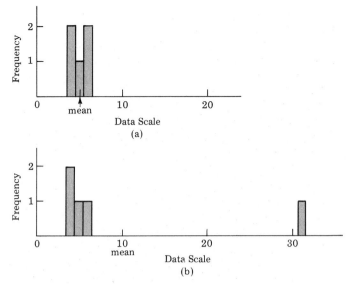

FIGURE 16    Histograms for Data Sets (6, 4, 5, 6, 4) and
(4, 4, 5, 6, 31)

multimodal data. In the latter instance we are interested in an index
that is typical of the data in general.

Some authors refer to the mean, median, and mode(s) as *indexes
of location*. "Central tendency" is more commonly used, but as we have
seen, the use of the word "central" could be confusing when data have a
multimodal or skewed distribution.

In some reports on the daily activity of the stock market the
average (mean) change in a share of stock is reported. Hence, if on the
New York Stock Exchange 11 million shares were traded with a mean
loss of 45¢, then the market value of the stocks was reduced $4.95
million on that particular day. Such a use of the mean is proper since
an indication of the total decrease or increase on the market is a good
indication of how the market as a whole, or in total, faired on that day.

We have already discussed an example of bimodal data, in the
graph in figure 13 concerning the years of school completed. The mean
is 11.8 years for the data set illustrated by figure 13. The median for
these data is 12.3 years. But neither of these indexes can by themselves properly describe the education of United States citizens, even
though the mean is the index all too often used to describe average
education. The proper indication of the average here would be the
modes. An indication that a major mode exists at 12 years with a minor
mode at 8 years is much more descriptive than either the mean or the
median of the data.

In summary, for proper interpretation of statistics designed to describe the central tendency of a data set, we must know the distribution of the data as well as the purpose for which the index is to be used. If we are interested in the data total, the mean is most descriptive. If we want an indication of the more typical or representative data values, the distribution of the data determines the index to use:

1. For unimodal and symmetric data or data that are nearly so use the mean.

2. For unimodal and skewed data use the median.

3. For multimodal data use the modes.

The type of data may also indicate the index of central tendency used as only the mode is defined for categorical data.

## Indexes of Dispersion

> *Indexes of dispersion* are statistics used to describe how data tend to be spread.

We are usually interested in the dispersion, or spread, of data about different central values. Indexes of dispersion are not used as widely as the indexes of central tendency just discussed, but they are important descriptive statistics. As we look at the range, standard deviation, and percentiles, we will see that the distribution of data plays an important role in a determination of which of these indexes of dispersion is most descriptive of the dispersion of the data. As we found with certain indexes of central tendency, particular indexes of dispersion can be of questionable descriptive value for skewed data.

### Range

The word "range" has several meanings. It may refer to the difference between the largest and smallest values in a data set. This number, coupled with some indication of central tendency, gives a limited indication of the dispersion of the scores in a data set. The word range is also used to refer to a pair of numbers, the smallest and largest numbers in a data set. In common usage the word range refers to a pair of numbers. We will consider the *range* to be either the data pair or the difference between these values.

The range is the index of dispersion that is easiest to calculate and is generally used by the news media. Consumers Union reports the mean and the range of the miles per gallon (mpg) obtained for a particular make of car. The range of mpg for a particular make of car can be descriptive since we might decide to buy a car based upon the reported range of gas mileage. (As with any statistic, however, such a feeling should be coupled with a consideration of the background influences on the data itself.)[7]

The weather report provides us with the range of temperatures for a given day with a forecast of the high and low temperatures for the coming day. This is interesting information, but it is limited. Two days with the same range of temperatures can be quite different, depending on what part of the day the temperature is near the high (or low) reading.

Figure 17 indicates the range of popularity of 5 presidents, as measured by a Gallup poll.

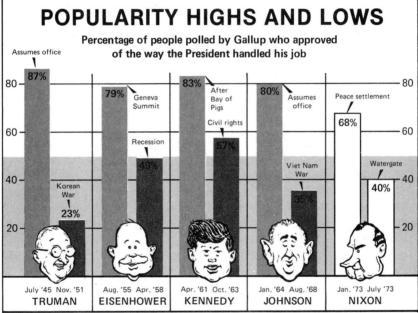

## POPULARITY HIGHS AND LOWS

Percentage of people polled by Gallup who approved
of the way the President handled his job

TIME Chart by J. Donovan and M. White

Presidents rise and fall in favor with a rhythm usually reflecting quite obvious crises and achievements. John Kennedy's chart is somewhat anomalous at first glance. His 83% peak of popularity just after the Bay of Pigs disaster represented a rally-round-the-President mood and soon dropped. His low in the autumn of 1963 resulted from a huge disaffection among white Southerners after the summer of the sit-in movement and the March on Washington.

**FIGURE 17**   Range of Popularity Levels for Presidents Truman, Eisenhower, Kennedy, Johnson, and Nixon (through July 1973)

Source: Reprinted by permission from TIME, The Weekly Newsmagazine; Copyright Time Inc., 1973.

We see that the range gives some indication of how the monthly popularity levels of the 5 presidents are spread. But many questions remain unanswered. For example, was the fall of President Johnson's popularity a gradual erosion as the Vietnam War dragged on? Did President Nixon lose popularity sharply as Senate Watergate hearings were aired over national television?

We might get a better idea of the information that knowledge of the range does *not* convey by looking at figure 18. Figure 18 shows the levels of President Nixon's popularity between his high point of 68% in January 1973 and his low point of 27% in February 1974. The range

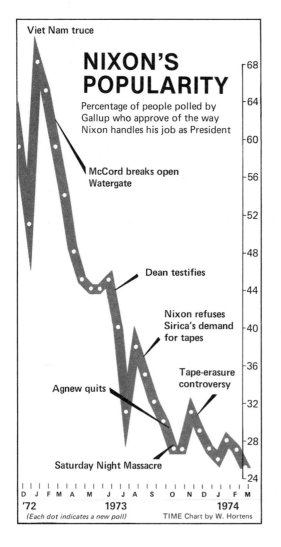

**FIGURE 18**
Level of the Popularity of President Nixon: December 1972 to March 1974

given in figure 17 is, therefore, a limited indication of dispersion of the monthly popularity levels of President Nixon.

We see that the range does describe dispersion to some extent. On the other hand, the range gives little indication of how the numbers or data are arranged between the largest and smallest data values. The range is therefore limited in the information it conveys about dispersion, but, nonetheless, the range is an important bit of information when coupled with other descriptive statistics.

### Variance and Standard Deviation

The variance and the standard deviation are other indexes of dispersion. Karl Pearson coined the term "standard deviation" in 1894; R. A. Fisher first used the word "variance" in 1948.

These statistics are most often used as inferential statistics but they have limited use as descriptive statistics. The standard deviation is the statistic of most descriptive interest but we must first calculate the variance before obtaining the value of the standard deviation for a set of data.

---

The *variance* is the sum of the squared distances of data from the mean divided by the number of data.

$$\text{variance} = \frac{\text{sum (difference of each datum from mean)}^2}{\text{number of items in data set}}$$

---

For example, the data 0, 6, 6, and 8 have a mean of 5

$$\text{mean} = \frac{0 + 6 + 6 + 8}{4} = \frac{20}{4} = 5$$

and a variance of 9.

$$\text{variance} = \frac{(0-5)^2 + (6-5)^2 + (6-5)^2 + (8-5)^2}{4}$$

$$= \frac{25 + 1 + 1 + 9}{4} = \frac{36}{4} = 9$$

We do not generally use the variance as an index of dispersion because it is in squared units. If we are talking about $0, $6, $6, and $8, the variance is 9 **squared dollars.** (What does a spread of 9 squared dollars mean?) Instead, we use the square root of the variance, called the standard deviation.

---

***Standard deviation*** = square root of variance

---

In our example

standard deviation = square root of 9 = 3 ***dollars***

The standard deviation is therefore a measure of how scores tend to be dispersed about the mean. This statistic is used more as an inferential statistic than as a descriptive statistic.

Some statistics texts define variance as the sum of the squared deviations about the mean divided by one less than the number of data (the degrees of freedom for the sum of squares of interest). This modification has nothing to do with the use of the variance as a descriptive statistic but affects its use as an inferential statistic. The interpretation of variance and standard deviation as a measure of dispersion is not appreciably affected by this difference in definition.

### Empirical Rule

We can interpret the standard deviation as a measure of dispersion by using the ***empirical rule,*** which describes the relationship between the mean and standard deviation for certain data sets. The rule is stated graphically for you in Figure 19. The empirical rule is meant as a crude "rule of thumb" for interpreting the standard deviation as a description of dispersion. The percentages in figure 19 come from data that are normally distributed. Normal data has a bell-shaped symmetric distribution, as in distributions *c, d,* and *e* of figure 14. The mean is used to describe central tendency for unimodal symmetric data even though the mean, median, and mode are equal. For a data set that follows closely to a bell-shaped distribution the mean, coupled with the standard deviation through the empirical rule, can be very descriptive.

### Percentiles

Since the percentages reported in the empirical rule hold remarkably well for data sets whose distribution does not deviate too much from being unimodal and symmetric, the empirical rule is often used to indicate how data are spread about the mean in such cases. Before illustrating when this rule might be unreliable, we must discuss percentiles, another method of indicating dispersion.

A *percentile* is a number below which are a certain percentage of the data in a data set.

For example, the 90th percentile is the number below which there are 90% of the numbers in a data set. The median is therefore the 50th percentile.

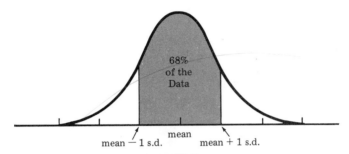

(a) *68%* of the data are within *one* standard deviation of the mean;

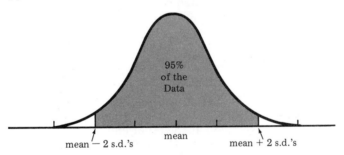

(b) *95%* of the data are within *two* standard deviations of the mean;

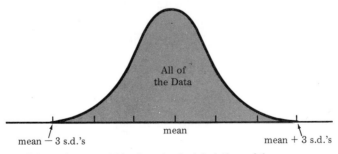

(c) *all* data values are within *three* standard deviations of the mean.

(s.d. = standard deviation)

**FIGURE 19    Graphic Representation of the Empirical Rule**

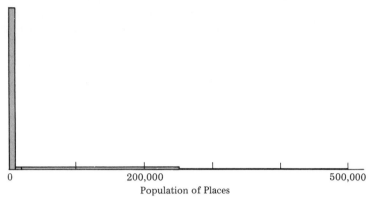

**FIGURE 20**   **Histogram of the Size of Places in the United States in 1970 (Places of population greater than 500,000 are not included in the graph.)**

Source: U.S. Department of Commerce, Bureau of the Census.

Let us consider certain percentiles relative to the size of places (incorporated and unincorporated concentrations of population) in the United States in 1970 (see figure 20). The 25th percentile is approximately 400 — that is, about 25% of the places had populations below 400. The median is about 1,135, and the 75th percentile is about 3,855. We see that the form of the distribution of the size of places is $L$-shaped. The data below the median of 1,135 are extremely clustered together, and the data, or size of places, above the median are greatly dispersed.

In figure 20, the dispersion at one end of the data scale is quite different from the dispersion at the other end of the data scale. This is true for any set of skewed data. The difference in dispersion from one end of the data scale to the other depends on the degree of skewness. A single index of dispersion such as the standard deviation cannot indicate varying amounts of dispersion within the data range. Indicating different percentiles can properly describe the dispersion of scores in such cases.

The empirical rule, as might be expected, fails to hold for this extremely skewed data set. Essentially all of the places are within one standard deviation, around 98,400, of the mean of 2,150.

Percentiles are also used when results of standardized tests such as the Graduate Record Exam are reported. An individual's test score and an indication of the percentage of lower scores are reported.

To summarize, indexes of dispersion are statistics that describe how numbers are spread. The range, the standard deviation, and percentiles are three examples of such indexes. These indexes give us a little more information about a set of data. Such information, when coupled with a graph showing the distribution of the data and some indication of central tendency, can give a very good description of a set of numbers.

## Describing a Data Set

Given descriptive statistics for a data set, how would you *describe* the data? Let us look at an example.

Consider the descriptive statistics in figure 21 based on a collection of faculty salaries from the arts and science college of a university in 1972. How could we describe these faculty salaries? Observing that the data have a bimodal distribution, we decide to use the two modes to describe central tendency. (We should think of this as clustering tendency, since there is no central tendency.) The major mode of $11,400 indicates a large clustering of salaries at this level. Faculty at the instructor or the assistant professor level would likely have salaries in this area of the data scale. The minor mode of $21,600 is indicative of a clustering of salaries for associate and full professors.

Dispersion can best be described using the percentiles. We observe that the range of the salaries, the 25th percentile, the median, and the 75th percentile divide the data into 4 equal parts — that is, one fourth of the salaries are between $6,300 and $9,800 (a difference of $3,500); one fourth of the salaries are between $9,800 and $12,700 (a difference of $2,900); one fourth of the salaries are between $12,700 and $16,000 (a difference of $3,300); and, one fourth of the salaries are between $16,000 and $30,000 (a difference of $14,000). Dispersion is of varying degrees throughout the range of salaries, with salaries varying most for the higher incomes.

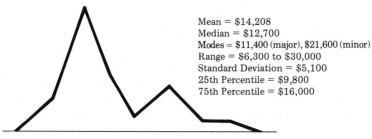

Mean = $14,208
Median = $12,700
Modes = $11,400 (major), $21,600 (minor)
Range = $6,300 to $30,000
Standard Deviation = $5,100
25th Percentile = $9,800
75th Percentile = $16,000

**FIGURE 21  Descriptive Statistics for Faculty Salaries Data**

## Descriptive Statistics: the Mechanics (Optional)

It might be instructive to show the details of how to describe a data set. Listed in table 3 is the electrical output (in megawatts) for a utility company for each weekday in July 1971. Observe that the numbers are discrete measurements made on the continuous variable of electrical output.

Graphic display of discrete or continuous data often requires grouping the data into classes. Such a grouping can give a concise picture of the data. But we must be careful. As illustrated in this chapter, it is easy to draw an improper "picture" of a data set.

The number of classes for a class frequency distribution is usually

2, if the number of data values is greater than $2 = 2^1$ but less than or equal to $4 = 2 \times 2 = 2^2$;

3, if the number of data values is greater than $4 = 2^2$ but less than or equal to $8 = 2 \times 2 \times 2 = 2^3$;

4, if the number of data values is greater than $8 = 2^3$ but less than or equal to $16 = 2 \times 2 \times 2 \times 2 = 2^4$;

5, if the number of data values is greater than $16 = 2^4$ but less than or equal to $32 = 2 \times 2 \times 2 \times 2 \times 2 = 2^5$; etc.

Since we have 22 data values and 22 is between 16 and $32 = 2^5$, we will use 5 classes. These 5 classes should be of sufficient size to cover the range of scores. Here the difference is

$1{,}273 - 889 = 384$
and    $384/5 = 76.8$ (about 80)

so we will use 5 classes of size 80. The smallest class must contain 889. There are 80 such classes including

810 to 889, 811 to 890, 812 to 891, . . ., 879 to 958, 880 to 959, etc.

We will use the class 880 to 959, since 880 is a multiple of the class size of 80, that is, 80 divides 880 evenly. Our classes are then

880 to 959, 960 to 1,039, 1,040 to 1,119, 1,120 to 1,199, and 1,200 to 1,279.

(Since 880 is a multiple of 80, all lower class limits are multiples of 80, that is, 960, 1,040, 1,120 and 1,200 are all evenly divisible by 80.)

The number of scores in each class, the frequency of each class, is then recorded in a *frequency table*, as shown in table 4. The frequency

**TABLE 3    Utility Data**

| Day | Output | Day | Output |
|---|---|---|---|
| 1 (Thursday) | 1156 | 16 | 1174 |
| 2 | 1073 | 19 | 1105 |
| 5 | 889 | 20 | 1066 |
| 6 | 1045 | 21 | 1062 |
| 7 | 1155 | 22 | 1078 |
| 8 | 1212 | 23 | 1126 |
| 9 | 1273 | 26 | 1179 |
| 12 | 1130 | 27 | 1068 |
| 13 | 1136 | 28 | 1063 |
| 14 | 1111 | 29 | 1081 |
| 15 | 1121 | 30 | 1061 |

Source: R. R. Hocking and L. R. La Motte, "Using the SELECT Program for Choosing Subset Regressions," *Proceedings of University of Kentucky Conference on Regression*, 1973.

table contains a listing of each class, a tally of the number of data in each class, the frequency or total tally for each class, the cumulative frequency, and percent cumulative frequency for each class. The cumulative frequency is the total number of data in a particular class **plus** the number of data in the lower classes. Therefore, since there are 12 data in the class 1,040 to 1,119 or in a lower class, the cumulative frequency for this class is 12; the class 960 to 1,039 has no data; the class 880 to 959 has 1 datum; and, the class 1,040 to 1,119 has a frequency of 11.

$$11 + 0 + 1 = 12$$

The percent cumulative frequency is the cumulative frequency divided by the total number of data values (22 in this case).

Since these measurements are continuous, they represent rounded values — that is, electrical outputs between 1,155.5 megawatts and 1,157.5 megawatts are recorded at 1,156 or 1,157. We usually round to the nearest unit of measurement. Hence, an electrical output between 1,156.5 and 1,157 is rounded up to 1,157, and outputs between 1,156 and 1,156.5 are rounded down to 1,156. If a datum value falls exactly at the halfway point, 1,156.5, let's say, we round to the

**TABLE 4    Data Summary (Utility Data)**

| Class | Tally | Frequency | Cumulative Frequency | Percent Cumulative Frequency |
|---|---|---|---|---|
| 880 to 959 | 1 | 1 | 1 | 4.5 |
| 960 to 1039 | | 0 | 1 | 4.5 |
| 1040 to 1119 | ͰͰͰ ͰͰͰ 1 | 11 | 12 | 54.5 |
| 1120 to 1199 | ͰͰͰ 111 | 8 | 20 | 90.0 |
| 1200 to 1279 | 11 | 2 | 22 | 100.0 |

even unit. Here, 1,156.5 would be rounded to 1,156. Similarly, a value of 1,157.5 would be rounded to 1,158, an even value.

Rounding should not cause too many problems in data description if it is properly carried out and if the discrete units used are properly calibrated; you would not want to round heights of people to the nearest yard or cattle weights to the nearest ton.

A graph in which the **midpoint** of each class, sometimes called the **class mark,** is represented on the horizontal scale and frequency is represented on the vertical scale is called the **frequency polygon.** (Recall that a polygon is a good way to graph continuous data.) Such a graph is illustrated in figure 22. The midpoint of the class is the mean of its endpoints. For example, the midpoint of the class 1,120 to 1,199 is

$$\frac{1,120 + 1,199}{2} = \frac{2,319}{2} = 1,159.5$$

The graph shown in figure 23 is the **ogive** (ō · jiv), which is a graph of percent cumulative frequency to the upper (larger) real limit of a class. The **real limits** of a class are the data values that were rounded to numbers in the class. The real limits of the class 1,040 to 1,119 would be 1,039.5 and 1,119.5 since continuous electrical outputs between these numbers were rounded to a data value between (or including) 1,040 and 1,119. (Recall that 1,040 and 1,119 are called the **class** limits.) The upper real limit of the class 1,040 to 1,119 is then 1,119.5.

The mean of the utility data is 1,107.5. This number can be approximated in data sets in which too many data make it impractical to add all the data together. (Even when one has access to a calculator,

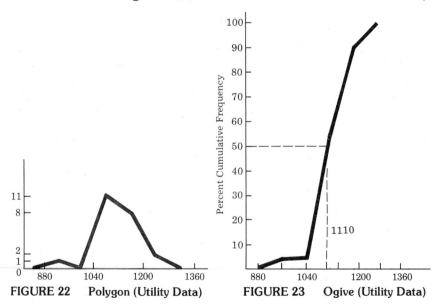

FIGURE 22   Polygon (Utility Data)

FIGURE 23   Ogive (Utility Data)

it may be desirable to find a quicker method of calculating the mean.) We can approximate the mean of data, which are summarized in a class frequency table, by dividing the sum of the products of the individual midpoints multiplied by their respective frequencies over all the classes by the number of data values. In our case we would approximate the mean by:

$$\frac{(919.5 \times 1) + (999.5 \times 0) + (1{,}079.5 \times 11) + (1{,}159.5 \times 8) + (1{,}239.5 \times 2)}{22} = 1115.9$$

The median of the 22 values is 1,108. This number can be approximated from the ogive, since the median has a percent cumulative frequency of 50. As seen on the ogive, the median is approximately 1,110.

We can also read other percentiles from the ogive. The 25th percentile of this data set is 1,080. Note the 25th percentile is the number with a relative cumulative frequency of 25. Likewise, the 75th percentile is about 1,170 — that is, about 75% of the electrical outputs listed in table 3 are below 1,170. (In fact, we see that 18 of the 22 data values, or 82%, are below 1,170.)

The range of these data is 889 to 1,273.

The standard deviation is found by first looking at the squared deviations of data from the mean, which are listed in table 5.

TABLE 5    Variance Calculation (Utility Data)

| Score | Score Minus the Mean | (Score Minus the Mean) Squared |
|-------|----------------------|-------------------------------|
| 1156 | 48.5 | 2352.25 |
| 1973 | −34.5 | 1190.25 |
| 889 | −218.5 | 47742.25 |
| 1045 | −62.5 | 3906.25 |
| 1155 | 47.5 | 2256.25 |
| 1213 | 105.5 | 11130.25 |
| 1273 | 165.5 | 27390.25 |
| 1130 | 22.5 | 506.25 |
| 1136 | 28.5 | 812.25 |
| 1111 | 3.5 | 12.25 |
| 1121 | 13.5 | 132.25 |
| 1174 | 66.5 | 4422.25 |
| 1105 | −2.5 | 6.25 |
| 1066 | −41.5 | 1122.25 |
| 1062 | −45.5 | 2070.25 |
| 1078 | −29.5 | 870.25 |
| 1126 | 18.5 | 342.25 |
| 1179 | 71.5 | 5112.25 |
| 1068 | −39.5 | 1560.25 |
| 1063 | −44.5 | 1980.25 |
| 1081 | −26.5 | 702.25 |
| 1061 | −46.5 | 2162.25 |
| Totals 24,365 | 0 | 118,431.50 |

TABLE 6    Approximating the Variance (Utility Data)

| I | II | III | IV |
|---|---|---|---|
| | | Squared | Column |
| | Midpoint | Deviation of | III |
| | Minus Approximate | Approximate Mean | Times the |
| Midpoints | Mean of 1115.9 | from Midpoints | Class Frequency |
| 919.5 | −196.4 | 38,572.96 | 38,572.96 |
| 999.5 | −116.4 | 13,548.96 | 0 |
| 1,079.5 | −36.4 | 1,324.96 | 14,574.56 |
| 1,159.5 | 43.6 | 1,900.96 | 15,207.68 |
| 1,239.5 | 123.6 | 15,276.96 | 30,553.92 |
| | | Total | 98,909.12 |

The variance is then

$$\text{variance} = \frac{118,431.5}{22} = 5,383.25$$

and

standard deviation = square root of $5,383.25 = 73.4$

As with the median and mean, when there are many data values, we might wish to approximate the variance by using the summarized data in the class frequency table. Table 6 illustrates the procedures.

The total of column IV is divided by 22 to get an approximate variance of 4,495.9. The approximate standard deviation is then the square root of this value, or 67.1.

# Summary

Chapter 3 has provided an overview of natural and descriptive statistics. At the outset of the chapter, natural, or census, statistics were discussed. In 1790 the United States became the first country to conduct a periodic census. Today the United States conducts a number of censuses including the Census of Housing, Census of Population, Census of Manufacturing, Census of Agriculture, Census of Mineral Industries, Census of Governments, Census of Unemployment, Census of Business, and the Census of Transportation. Only a fraction of the data collected by the Census Bureau comes from a complete enumeration of the population. Most data collected by the bureau comes from sampling the population of interest.

Having considered census statistics, we went on to discuss graphs as descriptive statistics. We must be careful to investigate the statistics that are being graphically displayed. A quick glance at a graph can lead to an improper interpretation of the statistical information.

An important consideration in this chapter was statistics known as measures of central tendency. When determining if a measure of central tendency such as the mean, median, or mode(s) is being properly used, we must consider both the distribution of the data and the use to be made of the description of the center of a data set. A consideration of the distribution of the data gives us an idea of the relative values of these statistics as well as their descriptive potential. If our interest is with the total of the data values, the mean is most descriptive. When interested in the most typical or representative value of a data set, we would use the mean, the median, or the mode(s), depending on the distribution of the data. For unimodal symmetric data use the mean; for unimodal skewed data use the median; and for multimodal data use the modes. For categorical data the mode(s) is the only index of central tendency that is available.

Dispersion of data about central values was another consideration of this chapter. We discussed the range, the standard deviation, and percentiles. The range is a limited descriptive statistic, since it gives little indication of the spread of data between the extreme data values. The dispersion of multimodal or skewed data sets is best described by percentiles, since dispersion will often vary at different parts of the score scale. For data sets that do not deviate too greatly from being unimodal and symmetric the standard deviation is descriptive with the understanding that:

68% of the data are within one standard deviation of the mean;

95% of the data are within two standard deviations of the mean;

Essentially all the data values are within three standard deviations of the mean.

We encounter many descriptive statistics in the media. Weather forecasters describe weather data, and someone always seems to be interested in the "average" member of some collection of people. Misuse of descriptive statistics can be detected if we visualize the distribution of the data being described. By knowing the distribution we can determine how to best describe central tendency and dispersion. However, it can be very difficult to visualize the distribution of data. We should expect that the media make this information available. As we become more knowledgeable about statistics, we will expect to be given more complete and more accurate statistical reports.

# No Comment

■ *"10 years in the life of the man on the go: Standing — 8600 hours waiting for appointments, Walking — 89 miles to and from parking lots, Waiting — 16,500 minutes in long lines, Running — 223 miles to catch a bus, train or plane, Rushing — 5336 hours to be on time, Stepping — on the accelerator 16,500 times and Stepping — on the brakes 72,000 times."*

*(From an advertisement for men's hose)*

# Exercises

1.  Verbally describe the data from which we calculated different descriptive statistics in the section entitled "Descriptive Statistics: the Mechanics." Any unusual aspects?

2.  Describe IQ scores that have a mean of 100 and a standard deviation of 15. For example, what proportion of people taking such a test would have scores between, let's say, 85 and 115? More than 130?

3.  (Optional) Data sets 1 and 2 were collected from a group of 24 students in a statistics class. (Data of this type might be collected from your class. Description of a more familiar set of data is likely to be of greater interest.)

    a. Data Set 1: Number of Marijuana Cigarettes Smoked in a Two-Week Period

    | | | | |
    |---|---|---|---|
    | 3  | 42 | 0  | 0 |
    | 0  | 0  | 0  | 0 |
    | 0  | 0  | 0  | 0 |
    | 0  | 14 | 50 | 0 |
    | 0  | 4  | 50 | 3 |
    | 0  | 0  | 3  | 4 |

    b. Data Set 2: Number of Alcoholic Drinks Consumed in a Two-Week Period

    | | | | |
    |---|---|---|---|
    | 25 | 0  | 3  | 4  |
    | 0  | 10 | 0  | 0  |
    | 16 | 0  | 0  | 2  |
    | 30 | 20 | 84 | 0  |
    | 4  | 15 | 24 | 12 |
    | 18 | 18 | 12 | 33 |

Calculate the different descriptive statistics mentioned in this chapter relative to data sets 1 and 2. Include a graph of these sets of data. Describe these data sets in your own words.

4. Suppose that a carcinogen (cancer-causing substance) is injected into mice and that the days to death from cancer are recorded. It is possible that some mice will not develop cancer. How would average time to death best be described — by mean days until death for mice that died or by median days to death of all mice injected with the carcinogen (assuming this number exists)? Explain.

5. Suppose the maintenance division of a university is seeking a contract for a five-year supply of light bulbs. Would you look for the company whose product has the greatest mean or the greatest median bulb life? Explain.

6. If you are buying tires for your car, would you prefer a tire brand that has the greatest mean or the greatest median miles of wear? Would the modes of multimodal wear data interest you? Explain.

7. Insurance companies set their rates according to the average size of claims made against them. Should such an average be expressed by a mean or median? Explain.

8. In a national news broadcast a commentator was describing (and comparing) the change in income of blacks and whites in the United States from 1960 to 1970. Should this commentator use the mean, the median, or the modal incomes of blacks and whites in 1960 and 1970. (Hint: What is the distribution of incomes?) Explain.

# For Discussion

1. Discuss possible background influences on the data given in Exercise 3 of this chapter. (Discuss the background influences on the data given or on a data set collected from your class.)

2. Census information is supposed to be confidential. Only area (block, census tract, county, etc.) summaries are reported. Dis-

cuss the need for privacy of such information in the execution of a credible census. Any potential misuses?

3.   The graphing technique shown in figure 24 is not too uncommon. Is it proper?

4.   Discuss the graph in figure 25. Is the variable "time" properly

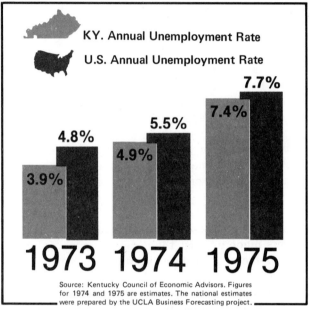

FIGURE 24    **Graph of Unemployment Statistics**

Source: Copyright © 1975, *The Courier-Journal and the Courier-Journal & Times.* Reprinted with permission.

FIGURE 25    **Decline of English Pound**

### Stock Market Plunge

Source: Reprinted by permission
from TIME, The Weekly Newsmagazine;
Copyright Time Inc., 1974.

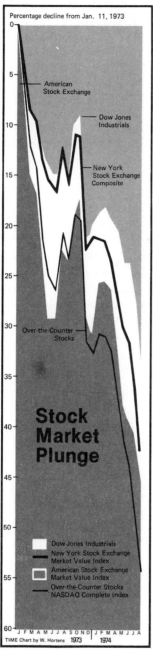

Percentage decline from Jan. 11, 1973

American Stock Exchange

Dow Jones Industrials

New York Stock Exchange Composite

Over-the-Counter Stocks

**Stock Market Plunge**

Dow Jones Industrials
New York Stock Exchange Market Value Index
American Stock Exchange Market Value Index
Over-the-Counter Stocks NASDAQ Complete Index

J F M A M J J A S O N D J F M A M J J A
TIME Chart by W. Hortens  1973  1974

Source: Reprinted by permission from TIME, The
Weekly Newsmagazine; Copyright Time Inc., 1974

graphed? Are there any problems involved with graphing dollars over the time period of 1924 to 1974?

5. Discuss the different possible distributions of daily rain levels for months with the same range.

6. What is an "average" American?

7. Discuss why government (national, state, or local) policymakers might be concerned with means (mean income, mean property tax, mean expenditures, etc.), and we citizens might be more interested in knowing medians or modes.

8. Discuss the technique used in the graph of stock market indexes shown in figure 26. What would be your impression of stock prices if the graph were drawn horizontally across the top of a page rather than vertically down the side of a page?

9. What types of data have we considered in this chapter? (Recall the types of data mentioned in chaper 2.) Any potentially important background influences on the different data sets we have considered?

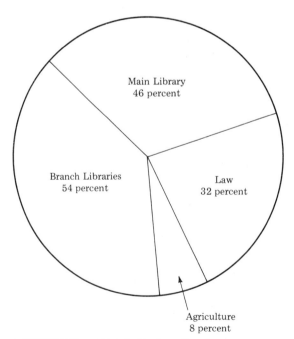

**FIGURE 27     Library Budget Graph**

10. The law requires that people answer census questions. What would be the effect on census data if this law were repealed?

11. The median age in this country has been increasing. (We alluded to this in the section "Index of Central Tendency" in this chapter.) Why do you think this is happening?

12. In describing the percentages of the total book budget going to different libraries on a university campus, the graph shown in figure 27 was used. Do you notice any unusual aspects in the graph?

# Further Readings

It is difficult to recommend other sources of information on descriptive statistics. The difficulty arises from the usual symbolic presentation of this topic, that is, you have to read "around" the mathematical formulas. If you would like to read on this topic, I suggest W. A. Wallis and H. V. Roberts, *Statistics: A New Approach* (New York: Macmillan, 1956).

Much has been written about the misuse of graphic statistics. You might look at any of the following: Campbell's *Flaws and Fallacies in Statistical Thinking,* Huff's *How to Lie with Statistics,* and Reichard's *The Figure Finaglers,* which were noted in the Further Readings section of chapter 2. For an interesting look at how graphs can be drawn to give a desired impression read the article "Grappling with Graphics" by Gene Zalagny *(Management Review* 64 [October 1975]: 4–16.

In the area of census data the Census Bureau has many readable reports. The bulletin listed in the Notes section would be helpful. Also consider P. M. Hauser's *Social Statistics in Use* cited in the Notes to this chapter for some interesting discussions in this area.

# Notes

1. H. H. Wolfenden, *Population Statistics and Their Compilation* (Chicago: University of Chicago Press, 1954), p. 4.
2. U.S. Bureau of the Census, *Fact Finder for the Nation* (1970): 1.
3. P. M. Hauser, *Social Statistics in Use* (New York: W. W. Norton, 1975). This book discusses the many uses made of census data.
4. J. S. Siegel, "Coverage of Population in the 1970 Census and Some Implications for Public Programs," *Bureau of Census Series* 56:2.
5. H. Arkin and R. R. Colton, *Graphs: How to Make and Use Them* (New York: Harper Brothers, 1936), p. 1.
6. R. S. Riechard, *The Figure Finaglers* (New York: McGraw-Hill, 1974) pp. 31–32.
7. *Consumer Reports* 39 (April 1974), p. 32.

# 4 Probability in Inferential Statistics

WE RECALL FROM OUR DISCUSSIONS in chapter 1 that we are often unable to investigate all of the population on which we desire information. We are often confronted with the problem of making inferences to a population after having studied only a part of, or a sample from, the population. There is uncertainty in such inductive inference. This uncertainty is measured by probability.

In this chapter we will first discuss the relative frequency interpretation of probability. Then we will consider the probabilities associated with the random selection of a number from a normal set of data. These probabilities will form the basis for a measurement of the uncertainty involved with inference to population characteristics. In this chapter we will introduce these important normal probabilities. In the next chapter we will see how these normal probabilities are tied to a particular inferential statistic, the sample mean. In chapter 6 we will discuss the inferential processes of estimation and hypothesis testing. It is at that point in our discussions that you will understand why we must, in this chapter, discuss relative frequency probability with special emphasis on normal probabilities.

As you read this chapter, keep the following in mind:

1. Probability has to do with the repetition of a random experiment. In statistical inference this random process is the selection of a random sample from a population.

2. To understand probability we must think about what happens when a random experiment is repeated many times; that is, to understand the uncertainty involved in statistical inference we must consider how a statistic changes in value from one random sample to another.

The next three chapters will be directed toward discussion of these concepts.

## Statistical Inference: Why We Need Probability

In chapter 1 we introduced the concept of inferential statistics. We discussed why it was not always possible to investigate the entire population of interest. You will recall it may be economically or practically unrealistic to attempt an investigation of an entire population. Manufacturers, as we mentioned, cannot afford to inspect every item they produce before making certain marketing and advertising decisions. Another reason why an entire population cannot be studied is that the population of interest may not exist at the time that information is desired. Preelection polling was given as one example of this situation. A third reason cited for not investigating an entire population was that the investigation could be destructive. Testing light bulbs, as we mentioned, requires burning them out. Clearly, a company cannot test all the light bulbs they produce before making marketing and advertising decisions.

Situations therefore exist in which there is interest in a population that cannot be investigated in its entirety. We are, however, able to study a part of, or a sample from, the population of interest. After investigating the sample, we must make determinations about the population. For example, after testing a sample of manufactured items we must make marketing and advertising decisions about the population of all the items the company produced. After talking to a sample of registered voters, a pollster will try to determine how all people are likely to vote on election day.

We must be aware, however, that the only way to be sure about the population of interest is to conduct a census of that population — that is, certainty is available only when we spend the time and money required for an investigation of the entire population. If we choose to investigate only a part of the population of interest, uncertainty enters the picture. Statements of fact concerning the population cannot be made if only a sample from the population has been studied. We must speak in terms of uncertainty when we speak about a population after having studied only a part of that population.

It would be desirous if we could select a sample in such a way that the uncertainty involved with making inference to population characteristics could be measured. And this can indeed be done; we can measure uncertainty with probability. In chapter 5 we will look at different sampling techniques for which we can measure the uncertainty of inference. But before we look at the measurement of the uncertainty in inferential statistics, let us investigate the measuring tool of probability.

## Probability: History and Theory

Prior to the development of probability theory as a science, which began in the 1600s, probability developed through its relationships with the insurance business and with games of chance. Marine insurance began in the fourteenth century in Italy and Holland. Since insurance companies needed to calculate the chance that they would have to pay a claim, probability became closely tied to such businesses. For example, around the year 1300 shipping of cargo required insurance premiums equal to from 12% to 15% of the value of the cargo, and intracontinental insurance rates were set at from 6% to 8%.[1]

Another use of probability arose in connection with the keeping of vital statistics on births, deaths, and marriages, beginning with plague death records kept during epidemics. Such a list, the **Bills of Mortality** in London, first appeared about 1517. These data gave rise to the calculation of the probability of death in a given time period as well as the probability of survival to a particular age. Such early analyses of these data were the beginnings of the life insurance business.

Most historians pay little attention to the development of probability as it relates to the insurance business; rather, they talk of the development of probability as it relates to games of chance. This might be due to the fact that evidence of games of chance go back to the beginning of recorded history. The first games of chance may have been the "rolling of the bones." The nearly symmetric heel bone, the astragalus (as·trag'·a·lus; plural, -li), of a hooved animal provided the chance mechanism for games much in the way that modern dice are the chance mechanism for many games today (see figure 1).

Excavations in Turkey have revealed use of the astragali of animals 13 centuries before Christ.

> On the inner side of the room were the ruins of a brick structure resembling a bar: behind it was two partly sunk clay vats, and in the corner a pile of ten beautiful chalices of the 'champagne' type. There was much other pottery in the room. Near the door was a pile of 77 knucklebones,

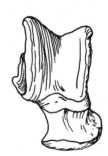

Sheep                                                      Dog

**FIGURE 1    Astragali**

Source Reproduced by permission of the publishers, Charles Griffin & Company Ltd of London and High Wycombe, from Florence N. David, *Games, Gods & Gambling.* 1962.

and beside them, perhaps used as tokens in a game, a stack of the crescent-shaped objects, till now usually described as loom-weights. The remainder of the floor space was mainly occupied by skeletons.[2]

Ancient Greek games involved the throwing of all 4 astragali of an animal, the "rolling of the bones." Modern empirical studies using the astragali of a sheep show the following probabilities of occurrences:

> .4 for the upper, broad and slightly convex, side (counted as "4" in classical games)
>
> .4 for the opposite, broad and slightly concave, side (counted as "3")
>
> .1 for the lateral, flat and narrow, side (counted as "1")
>
> .1 for the opposite, narrow and slightly hollow, side (counted as "6").[3]

A favorite research of scholars of the Italian Renaissance was to deduce the scoring of ancient games. The values mentioned above, 1, 3, 4, and 6, are a generally accepted scoring method.

The development of probability theory is closely tied to games of chance. In order to have the "upper hand" in games, professional gamblers sought and paid for the services of probabilists. Probability as it related to statistics developed through a similar relationship between probabilists and insurance companies.

### What is Probability?

The use of words such as "uncertainty," "chance," "odds," and "probability" has become extremely widespread. We are continually hearing statements like: "The probability of rain is 20% today, 30% tonight, and 60% tomorrow;" "The chances that the Cubs will win the pennant are very small;" or, "The odds in favor of Secretariat winning the Triple Crown are better than they have been for any other horse in the last 25 years."

Probability has even entered the courtroom. An example comes from a Swedish trial. A policeman had noted the position of the valve stems of the wheels on one side of a car much as a pilot notes direction: one valve at, let's say, 2 o'clock and the other at 6 o'clock. To the charge of overtime parking the defendant claimed he left the parking space and returned. It was chance, he claimed, that led to the valves returning to the same location. An expert testified that the chances of such a chance occurence is 1 in 144 (only 1 of the 12 × 12 = 144 possible outcomes favored the event; 12 represents the 12 clock positions). The judge ruled for the defendant saying that had all 4 wheels been observed by the policeman the coincidence claim of the defendant would have been rejected. ($12 \times 12 \times 12 \times 12 = 20{,}736$ or the odds of the chance occurrence of all 4 valves returning to the same positions is 20,736 to 1.)[4]

Probability is, as we've discussed, an integral part of inferential statistics. But what is probability and exactly how does it fit into our discussion of statistics? "It is unanimously agreed that statistics depends somehow on probability. But, as to what probability is and how it is connected with statistics, there has seldom been such complete disagreement and breakdown of communication since the Tower of Babel. There must be dozens of different interpretations of probability defended by living authorities."[5] (The Book of Genesis relates how the descendents of Noah built a high tower in order to make themselves a name and not be scattered over the earth. The Lord came to this people who spoke a single language and confounded their speech so they could not understand each other; hence, the reference to the Tower of Babel and the Biblical origin of different languages.)

Comments concerning the meaning of probability statements sometimes appear in the media. The host of a radio program, for example, explained to a caller what is meant by the statement, "The probability of rain is 50% today." His interpretation of probability was, "To say the probability of rain was 50% was to say that 50% of the area covered by the forecast would get rain." As we quoted, "There must be dozens of different interpretations," but this is yet another one.

Probability enters our discussion of statistics through the uncertainty of inferring from a sample to a population. An understanding of probability is therefore important if we are to be able to interpret inferential statistics. As we've seen, there are many ways to view probability. However, we will discuss only one interpretation of probability — *relative frequency probability*.

### Relative Frequency Probability and Random Experiments

To begin our discussions of relative frequency probability let us consider the probability that the flip of a coin will result in a head. Most of us would say that this probability is one half. Buy why do we feel that the probability of heads is one half? Let us look at the relative frequency interpretation of this probability statement: This view of probability is likely to explain why we feel the probability that a flip of a coin will result in heads is 1/2.

Basic to the relative frequency interpretation of probability is the performance of a random experiment.

> A *random experiment* is an experiment for which all possible outcomes are known but the outcome of a particular execution of the experiment cannot be predicted with certainty.

In our example the random experiment is the flipping of a coin. This is a random experiment since we cannot predict with certainty whether the result of our flip will be heads or tails, although it is assumed that one of these two outcomes must occur.

Before we define probability we must also understand what is meant by an event.

> An *event* is a collection of possible outcomes of an experiment.

The event of interest to us in the coin example is the single outcome "heads." The *probability* of an event is a number between zero and one. In the *relative frequency sense* the probability of an event is approximately the number of occurrences of the event divided by the number of times the experiment is run if we run the experiment "a large number of times."

$$\text{probability} = \frac{\text{number of occurrences of event of interest}}{\text{number of times experiment is run}}$$

Translated to our example, the probability of getting heads in the flip of a coin is 1/2 because if we flipped this coin "a large number of times," about 1/2 of the flips would result in heads.

As another example, consider the rolling of a die. The probability of rolling the face containing 6 dots is 1/6, as about 1/6 of the times that we roll the die a 6 would come up.

Interpreting the statement, "The probability of rain is 50%," in the relative frequency sense would be to say that if the present atmospheric conditions were to reappear a large number of times, it would rain about one half of those times. In this case our "experiment" would be the combination of different atmospheric conditions. The outcomes of interest would be rain or no rain. Again, this is a random experiment as we cannot predict with certainty whether a particular set of atmospheric conditions will lead to rain.[6]

In the last section we stated probabilities associated with throwing the astragalus of a sheep. If the heel bone of a sheep were thrown, 1 of each of the 2 broad sides would appear about 4 out of 10 times and each of the 2 narrower sides would appear 1 in 10 times. How are these probabilities likely to have been determined? Roll an astragalus a large number of times. The 4 sides appear in a ratio of about 4:4:1:1.

The probability that any of the 6 sides will appear in a roll of a modern die is 1/6. Does this mean that in 600 rolls of a die 100 1s, 100 2s, 100 3s, 100 4s, 100 5s, and 100 6s will appear? For that matter, if we flip a coin 1,000 times, will heads appear 500 times? The answers to both these questions is no. If a coin is flipped 1,000 times, 490 heads might appear; the next time we flip the same coin 1,000 times 515 heads might appear. To say that the probability of heads in the flip of a coin is 1/2 is to say that one expects to get 500 heads in 1,000 flips. In practice we will get *about* 500 heads in 1,000 flips. The actual number of heads in 1,000 flips of a coin varies about 500 in a predictable manner. We will return to this idea in chapter 5.

## The Birthday Problem (Optional)

In a room of 30 people there is a high probability that at least 2 people have the same birthday. Given that there are 365 distinct possibilities for a person's birthday, multiple birthdays within a group of only 30 individuals seems an unlikely event. Here is a case in which the probability of an event is not at all intuitive.

Let us look at the probability of multiple birthdays. Suppose we start our discussion by looking at the chances that at least 2 of 3 people have the same birthday. We will consider the *opposite*, or *complementary*, event of multiple birthdays. The complementary event is that all individuals have different birthdays — that is, the complement of the event that at least 2 of 3 people have a common birthday is that all 3 have different birthdays. Likewise, the complement of the event that at least 2 of 20 people have a common birthday is that all 20 people have different birthdays. The probability of an event and its complement are related in the following way: The probability of an event plus the probability of the complementary event is *always* 1.

Let us find the probability in question by looking at the probability of the complementary event — that is, what are the chances that 3 people have *different* birthdays? This probability subtracted from 1 will give us the probability that at least 2 of 3 people have a common birthday.

Given the birthday of the first of these 3 persons, a second person will have a different birthday 364 times out of 365, or with probability 364/365. We see this by observing that the second person's birthday is different from the first person's birthday if the second's birthday is one of the 364 days in a year that are *not* the first person's birthday.

Likewise, the third person will have a birthday distinct from the first 2 people with probability 363/365. The probability that 3 people have different birthdays is

$$\frac{364}{365} \cdot \frac{363}{365} = .992$$

Think of this as

$$\frac{365}{365} \cdot \frac{364}{365} \cdot \frac{363}{365} = .992$$

so that we don't lose track of the fact that we are talking about *3* people.

We assume that people are selected randomly and the occurrences of birthdays are independent events. Independent events have occurrences that are not related. For example, successive flips of a coin are independent events. The outcome of any one flip is not affected by the outcome of other flips of the coin. Similarly, the timing of a person's birth is assumed unrelated to the timing of the birth of another person.

The probabilities of independent events multiply. The probability that at least 2 of 3 people have a common birthday is then

$$1 - .992 = .008 \text{ or } 8 \text{ in } 1,000$$

Recall that the probability that 3 people have different birthdays plus the probability that at least 2 of these 3 people have a common birthday — the complementary event — is 1, since the sum of the probability of an event and the probability of the complementary event is always 1.

Similarly, the probability that at least 2 of 4 people have a common birthday is

$$1 - \frac{365}{365} \cdot \frac{364}{365} \cdot \frac{363}{365} \cdot \frac{362}{365} = 1 - .984$$

$$= .016 \text{ or } 16 \text{ in } 1,000$$

Table 1 gives the probability of multiple birthdays in groups from 3 to 40 people. We observe in table 1 that in a group of 23 people the chances of at least 2 having the same birthday is better than 50-50 (50.7%, in fact). For our original problem of whether at least 2 of 30 people have a common birthday, the chances are 7 out of 10 that at least 2 do have the same birthday.

TABLE 1    Probabilities of Multiple Birthdays

| Number of People | Probability at Least Two Have Same Birthday | Number of People | Probability at Least Two Have Same Birthday |
|---|---|---|---|
| 3 | .008 | 22 | .476 |
| 4 | .016 | 23 | .507 |
| 5 | .027 | 24 | .538 |
| 6 | .040 | 25 | .569 |
| 7 | .056 | 26 | .598 |
| 8 | .074 | 27 | .627 |
| 9 | .095 | 28 | .654 |
| 10 | .117 | 29 | .681 |
| 11 | .141 | 30 | .706 |
| 12 | .167 | 31 | .730 |
| 13 | .194 | 32 | .753 |
| 14 | .223 | 33 | .775 |
| 15 | .253 | 34 | .795 |
| 16 | .284 | 35 | .814 |
| 17 | .315 | 36 | .832 |
| 18 | .347 | 37 | .850 |
| 19 | .379 | 38 | .864 |
| 20 | .411 | 39 | .878 |
| 21 | .444 | 40 | .891 |

# The Gambler's Fallacy

As mentioned earlier in this chapter, the development of probability theory is due to a large extent to the analysis of games of chance. Particularly in the sixteenth and seventeenth centuries, a close relationship between gamblers and probabilists contributed greatly to the development of probability theory, which in turn was to influence the development of statistical theory.

One of the interesting aspects of the analysis of games of chance is the so-called "gambler's fallacy" that still plagues the amateur gambler. The situation might best be explained by reconsidering the flipping of a coin. Suppose a series of heads has occurred, let's say, 10 heads in the first 10 flips of a coin. (A series of "reds" in roulette or "passes" in craps are similar examples.) Given a series of 10 heads in succession, how would you bet on the eleventh flip of this coin? Would you give even odds that the eleventh flip will result in heads? Would you bet heavily on a tail (or a head) on the eleventh flip? An analysis of this problem should help us better understand probability as we discussed it in the last section.

The gambler's fallacy is the mistaken belief that a tail is more likely to occur on the eleventh flip of a coin after a series of 10 heads.

People generally feel that the number of heads and tails will even out eventually — that is, about one half of the flips of a coin will result in heads. The key to understanding the fallacy of the assumption that tails is more likely after 10 successive heads is the word "eventually." At some time in a continued series of flips of a coin, equilibrium (about one-half heads) should be reached. But must there be exactly one-half heads in 20 flips or exactly one-half heads in 100 flips of a coin? Not necessarily. Equilibrium need not occur at a particular number of trials — just "in the long run" or "eventually" (see chapter 5).

Another way to see the fallacy of betting heavily on tails after 10 successive heads is to realize that the coin has no way of remembering that it came up heads 10 times in a row, that is, the outcomes of successive executions of the random experiment of flipping a coin are independent. How is the coin to know it is to come up tails so things might even out?

With a relative frequency interpretation of the statement, "The probability of heads in the flip of a coin is one half," we would give even odds on the events of heads and tails on the eleventh flip of a coin, even after 10 successive heads. Therefore, when thinking of probability, be careful. If the probability of an event is, let's say, one half, we cannot expect that one half of any number of runs of the experiment will result in the event of interest. Repeated repetitions of the experiment will eventually lead to equilibrium.

In reality a coin is not likely to be exactly balanced so the probability of heads is different from 1/2. Under this assumption 10 successive heads implies the probability of heads might be greater than 1/2, indicating a bet on heads on the eleventh throw. This idea is closely tied to testing the hypothesis that a coin is fair (see chapter 6).

## The Normal Distribution and the Empirical Rule

Much of the probability inherent in inferential statistics is based on a *normal distribution* or normal curve. First reference to the normal curve is due to Abraham de Moivre (see chapter 1) with his 1733 publication of a second supplement to his *Miscellanea Analytica*.[7] De Moivre reported that repeated measurements of the same physical quantity behaved with surprising regularity. The pattern or distribution of these measurements was found to be closely approximated by the "normal curve." Figure 2 is a representation of data with a normal distribution.

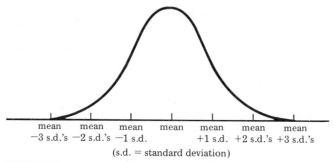

FIGURE 2    **A Normal Curve**

The fact that the normal distribution was found to occur in the physical world led to the extensive examination of this distribution by de Moivre, Pierre-Simon de Laplace, and Carl Friedrich Gauss. Gauss' work is especially noted. A normal distribution is often referred to as a Gaussian distribution.

Today the normal distribution plays a dominant role in inferential statistics. Besides its importance as an approximation of measurement errors, the normal distribution also gains importance in inferential statistics due to the *Central Limit Theorem.* Roughly speaking, this theorem states that many inferential statistics show surprising regularity in their variation from random sample to random sample. And this variation from sample to sample is closely approximated by the normal curve (see chapter 5).

The normal curve is extensively used in standardized testing, particularly in intelligence (IQ) testing. Such test scores are normalized, or transformed, to scores that approximate a normal distribution. Some teachers will even normalize classroom test scores in order to determine students' grades.

Let us look at this very important collection of numbers — the normal population, a population of numbers having a normal distribution. Only continuous data can have a normal distribution. Finite sets of data do, however, have frequency polygons that can be closely approximated by a normal distribution. We refer to such a finite data set as a *normal data set.* If we can count the number of data in a set of numbers, the data set is said to be *finite.* Recall distributions *c, d,* and *e* in figure 1 of chapter 3. These data had a unimodal symmetric distribution and, hence, represent nearly normal data sets.

In a normal distribution only two numbers are of interest, the mean and the standard deviation. Since the distribution of normal data is bell-shaped and symmetric, the mean, median, and mode coincide and are at the center of the curve (the point of symmetry).

As indicated in our discussion in chapter 3 of the empirical rule, the relationship between the mean and standard deviation is crucial to understanding a set of data that has a normal distribution. Recall that we stated that

68% of the data are within one standard deviation of the mean;

95% of the data are within two standard deviations of the mean;

All of the data are within three standard deviations of the mean.

These percentages, first determined by de Moivre in the eighteenth century, hold for normal data. In fact, we can be even more precise, as shown by the statement in table 2.

The probabilities given in table 2 can be interpreted using the graph in figure 3. For example, we see that about 90% (= B) of the numbers in a normal data set are within 1.645 (= A) standard deviations of the mean, and 87% (= B) of data with a normal distribution are within 1.5 (= A) standard deviations of the mean. (These probabilities are all given in Appendix B, table 1 at the end of this book for easier reference.)

**TABLE 2**    **Normal Probabilities***

| A | B |
|---|---|
| 3 | 99.7 |
| 2.5 | 99 |
| 2 | 95 |
| 1.645 | 90 |
| 1.5 | 87 |
| 1 | 68 |
| .5 | 38 |

*Within (A) standard deviation(s) of the mean are (B)% of data with a normal distribution.

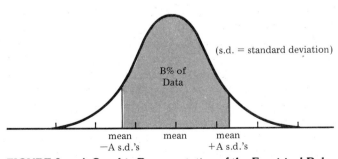

(s.d. = standard deviation)

B% of
Data

mean       mean       mean
−A s.d.'s             +A s.d.'s

**FIGURE 3**    **A Graphic Representation of the Empirical Rule**

We can think of the percentages expressed in our example as probabilities. Envision the numbers in a normal data set written on chips of wood and placed in a large drum. Suppose that this drum is turned so that the chips are thoroughly mixed. One chip is then selected from the drum. (Note that this mixing and selection process constitutes a random experiment as discussed earlier in this chapter.) The probability that this number is within two standard deviations of the normal mean is .95. More realistically, suppose that scores on an IQ test form a normal data set with a mean of 100 and a standard deviation of 15. What is the probability that a student selected randomly will have an IQ of between 70 and 130? An IQ of greater than 130?

Let us look at two examples of sets of normal data. Consider the graphs in figure 4(a) and figure 4(b).

The bell-shaped graph of the normal distribution can be interpreted much as the frequency polygons discussed in chapter 3. The graph gives an indication of the proportion of data that falls in various areas of the data scale. Probabilities associated with intervals, and only with intervals, on the data scale may be found from such a graph.

For the data graphed in figure 4(a), a graph of data like the IQ scores just mentioned, we would expect to find 68% of the data between

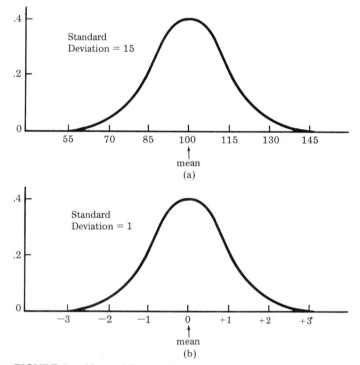

**FIGURE 4    Normal Distributions**

85 and 115. Ninety-five percent of the data in this collection would be between 70 and 130 — that is, if these were scores on a particular IQ test, 95% of people taking such a test would have IQs between 70 and 130. (Can you see that 2.5% of the data is above 130?)

Similar statements can be made about the data graphed in figure 4(b). The normal data graphed in figure 4(b) are called **standard normal.** The standard normal distribution has a mean of 0 and a standard deviation of 1. The absolute magnitude of a datum indicates the number of standard deviations above or below the mean the datum is. For example, the datum $-1.5$ is one and one-half standard deviations **below** the mean, and the datum $+2$ is two standard deviations **above** the mean.

We will discuss normal data again in chapter 5. At that time we will see how normal probabilities, as given in table 1, give us a measurement of the uncertainty involved with statistical inference.

## Two Distinct Groups: The Dichotomous Population

When dealing with categorical data, a nonnormal population arises. (Recall our discussion of types of data in chapter 2.) We will restrict our attention to data in two categories such as "yes"–"no," "male"–"female," "success"–"failure." In situations of more than two categories we might restrict our attention to a particular category, let's say, "for Candidate A" in an election. All other possible categories are grouped together into a category of "not for Candidate A." (This second category might include voters for a number of other candidates as well as undecided voters.)

Let us call the categories 1 and 0 in which 1 represents "success" and 0 represents "failure." When considering 1s and 0s we are interested in the proportion of 1s. For example: What is the proportion of voters "for Candidate A"? (1 represents a person "for Candidate A.") What proportion of products manufactured by a company are "defective"? (1 represents a "defective" item in this case.)

### Mean

The sum of all the 0s and 1s in our **dichotomous** (two distinct groups) **population** equals the number of 1s in our population. The mean is then equal to the proportion of 1s. For example, consider the set of data

0, 0, 1, 0, 1, 1, 0, 1, 0, 0

Such a set of data might represent 4 people who are "for Candidate A" (that is, there are 4 1s and 6 people who are "not for Candidate A").

$$\text{mean} = \frac{0 + 0 + 1 + 0 + 1 + 1 + 0 + 1 + 0 + 0}{10}$$

$$= \frac{4 \ (1s) \ + \ 6 \ (0s)}{10} = \frac{4}{10}$$

$$= \frac{\text{number of 1s}}{\text{total number in data set}} = \text{proportion of 1s in data set}$$

### Standard Deviation

The standard deviation of the 0s and 1s in a dichotomous population depends on the proportion of 1s. A graph showing the standard deviation for various collections of 0s and 1s including the standard deviation of a dichotomous population is given in figure 5. For example, suppose we have a collection of 0s and 1s in which 36% (.36) are 1s. This could be representative of a population of voters in which 36% are "for Candidate A" and the remaining 64% undecided or for other candidates. The standard deviation of this collection of numbers is about .5 as seen in figure 5.

The data set

0, 0, 1, 0, 1, 1, 0, 1, 0, 0

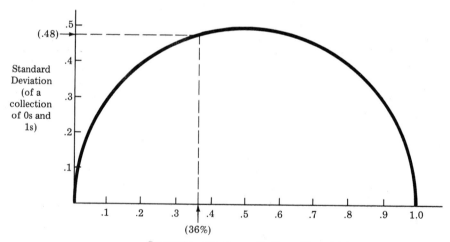

Proportion of 1s (in a collection of 0s and 1s)

**FIGURE 5**   Standard Deviation of a Collection of 0s and 1s

has .4 1s. The standard deviation is then .49. From chapter 3 recall that the variance of these data is

$$\text{variance} = \frac{6(0 - .4)^2 + 4(1 - .4)^2}{10} = .24$$

and

$$\text{standard deviation} = \text{square root of } .24 = .49$$

As for the data in this example, figure 5 will give the standard deviation of any collection of 0s and 1s. We must first determine the proportion of 1s in the collection and locate this value on the horizontal scale of figure 5. We then read the standard deviation of the collection of 0s and 1s on the vertical scale as indicated in figure 5 for a proportion of .36. (Figure 5 is also given in Appendix C at the end of the book, for convenience in future references to this graph.)

Similarly, if 1/2 of the data in a dichotomous population are 1s, then the standard deviation is also equal to 1/2. (Can you relate this population to flipping a fair coin?) If 25% are 1s, the standard deviation is .43, etc.

In chapter 5 we will see how the normal probabilities discussed in the last section can be applied to measuring uncertainty when sampling is from a dichotomous population. For example, suppose a sample is selected from a dichotomous population. A measurement of the uncertainty involved with inferring from the sample proportion of 1s to the population proportion of 1s can be approximated by normal probabilities!

## The Chi-Square Distribution

There are sets of continuous data that are not normally distributed. Time would not be judiciously spent going over all of the important distributions. We will, however, find need to look at one distribution besides the normal. This is the chi-square distribution. This distribution is not of interest because it occurs naturally as does the normal distribution. Recall that measurements of the same physical quantity tend to have a normal distribution. The chi-square distribution is important because the square of standard normal measurements has a chi-square distribution. This distribution derives its importance, then, from the Central Limit Theorem just as the normal distribution does. This distribution arises from analysis of categorical data as we will see in chapter 9.

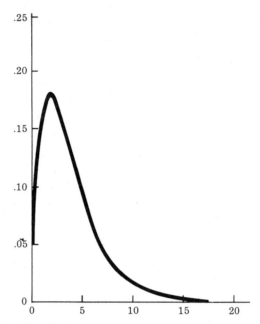

**FIGURE 6** The Chi-Square Distribution (degrees of freedom = 4)

### Degrees of Freedom

A set of continuous data with a chi-square distribution is skewed to the right. As mentioned earlier in this chapter, the normal distribution is characterized by two numbers, this mean and the standard deviation. The chi-square distribution is characterized by a single value called the *degree of freedom* for the distribution. The graph of chi-square distribution for degrees of freedom equal to 4 is given in figure 6.

Just as no finite set of data can have a normal distribution, no finite data set can have a chi-square distribution. Your intuition will, however, be aided by thinking of a finite set of data that has a distribution closely approximated by a chi-square distribution.

### Finding Chi-Square Probabilities

In our discussions the use of the chi-square distribution will be limited to a look at certain specific probabilities associated with such a data set. As in the normal case, we can visualize such probability by considering a random experiment. This experiment involves writing the data values with a chi-square distribution on wooden chips. One chip is drawn at random from the collection of chips. We ask the probability that the number drawn will exceed specified values. For example, for a set of data with a chi-square distribution with 1 degree of freedom we ask the probability of selecting a number larger than 3.8. Appendix B, table 2 gives us this probability, which is .05.

The 4 columns found in table 2 of Appendix B are the degrees of freedom column and the .1, .05, and .01 probability columns. The values in the probability columns are the numbers in a chi-square data set that will be exceeded with probabilities .1, .05, and .01, respectively. For example, suppose that data with a chi-square (1) distribution (a chi-square set of data with one degree of freedom) are put in a drum and mixed. If one number is selected, it will exceed 2.7 .1 = 10% of the time. Similarly, a chi-square (1) value will exceed 3.8 and 6.6 with probabilities .05 and .01, respectively.

Similarly, using table 2 we can see that there is a .01 probability of randomly selecting a chi-square (4) value that is larger than 13.3. What is the chance of selecting a chi-square (7) value above 14.1? (Do you see that this probability is .05?)

In chapter 9 we will see how the chi-square probabilities given in table 2 of Appendix B measure the uncertainty in particular examples of statistical inference involving categorical data.

## Decision Making

Let us investigate the decision-making process by looking at an example. We seek a better understanding of why we make the decisions we do as well as improved comprehension of why others, notably politicians, make the decisions they do. The decisions of the latter can have a profound effect on our lives and should therefore be subjected to careful analysis.

As a means of analyzing the essential aspects of the decision-making process let us consider an application familiar to all of us, at least indirectly. Suppose that we are sitting on the jury of a criminal trial and are confronted with the problem of determining whether the person accused of a crime is guilty or not guilty.

### Prior Feelings

Before the presentation of evidence we are likely to have some feelings about the guilt of the defendant. Although our judicial system requires that we consider the defendant not guilty until proved guilty, we are likely to have some prior feelings about the guilt of the defendant.

A major effect on our prior feelings is pretrial publicity. With the great amount of media coverage of crime it is hard to believe that anyone who reads the newspaper or listens to the radio or television does not have some prior feelings about the guilt of a defendant, especially with respect to crimes of high public interest.

We see, then, that the first influence on our decision-making process is the personal beliefs we have about the question at hand, prior to any investigation we might carry out or might be carried out by others.

## Types of Errors

Realizing that we could make an error in judgment, our concern turns to the consequences of making an incorrect decision. There are two types of error possible in the jury example. The first is traditionally called an error of the first kind or *Type-I error,* since it was first investigated in detail. A Type-I error would be to vote guilty when in fact the defendant was not guilty. The consequences of making a Type-I error would be to send a person to jail for a crime he or she did not commit.

A second type of error, a *Type-II error,* would occur if we declared the defendant not guilty when he or she was in fact guilty of the crime in question. The consequence of committing an error of this kind would be to return a criminal to society.

Each of us has a feeling about which of these two types of error we consider most worth guarding against. The particular crime in question would be one of the many influences on such a determination.

It should be noted that a fear of making a Type-I error (sending an innocent person to jail) is what led our judicial system to require unanimity of a jury for conviction. If all members of a jury must vote guilty, fewer defendants are found guilty and so fewer innocent people are put in jail. Strong feelings against returning criminals to society, that is, strong feelings against making a Type-II error, are leading some states to change the requirement that all jurors must vote guilty before a guilty verdict is rendered. Some states now require 8 or 10 of 12 votes for a guilty verdict.

## Interpretation of Information

Let us return to the trial for the presentation of evidence in defense and in prosecution of the defendant. We must weigh the sometimes conflicting testimony, or, as we shall say later, we must *interpret the information presented to us.* The same evidence is likely to be interpreted in as many different ways as there are jurors. During deliberation we jurors will discuss and defend our views and question the views of others. In the end each must decide.

So we see that there are three influences on the decisions people make: prior feelings, consideration of the consequences of making an incorrect decision, and interpretation of available information. It is hard to assess the effect each influence will have on a particular decision that someone makes. These influences will not be the same for different people or for different decisions made by the same person. Nevertheless, since these factors *do* influence the decisions people make, we must consider these influences if we are to better understand the decisions we make as well as the decisions made by others.

### Influences on the Decisions of Others

Many decisions made by others greatly affect our lives. Most notable are the decisions made by our elected officials. As a second example of the influences on decision making, consider a senator's decision to vote for or against a particular piece of legislation. Just as in most decision-making processes, an integral part is played by prior feelings, usually reflecting a senator's conservative or liberal leanings. Senators must also consider the consequence of voting for a bill that their constituencies and/or congressional colleagues oppose. The most interesting aspect of decisions made by elected officials is that 100 individuals (in the case of the United States Senate) seem to interpret information on pending legislation in 100 different ways. These interpretations are influenced both by prior beliefs and by feelings about the consequences of making an error in judgment; thus, the three influences on the making of a decision are interrelated and are nearly impossible to separate.

# Decision Theory: An Example (Optional)

With the examples of the last section still in mind, let us consider the very perplexing question, "How might we measure the components of the decision-making process?"

First, could you quantify your prior feelings concerning a defendant's guilt by declaring that the prior probability of guilt is, let's say, 20%? Could a senator quantify his prior feelings of how beneficial a proposed piece of legislation might be? (Such a determination is analogous to setting a bet. If a person gives 2 to 1 odds in a bet that a particular team will win a football game, he or she is setting the probability that this team will win to be 2/3.)

It is not likely that a person will quantify his or her prior feelings before making a decision. This does not necessarily mean that prior feelings do not influence the decision made. Prior feelings influence many of the decisions people make whether or not one attempts to quantify this influence.

Second, could we quantify our fear of making an incorrect decision. If a Type-I error is one unit, then can you comparatively quantify your feelings about making a Type-II error? For example, assuming that you feel that the cost of putting an innocent person in jail is one unit, how many units would you place on your desire not to put a criminal back into society? Could a politician quantify his or her feelings about alienating party leaders and/or constituents with an ill-conceived vote?

If the loss incurred from making an incorrect decision is financial, quantification is possible. However, in many of the decisions we make we cannot quantify the loss that we would incur if we made an incorrect decision. The decisions we make are nonetheless influenced by our fear of making a wrong decision.

The third component of the decision-making process is the interpretation of the information we collect to aid us in making a good decision. This information may or may not be statistical.

Let us look at a particular decision problem to help illustrate how we might proceed if we could indeed quantify the components of a decision problem. We will consider a farmer's decision on what to plant. Suppose a farmer has a section of land on which he could plant corn, soybeans, or alfalfa. No matter which crop the farmer decides to plant, the weather could affect his yield. He therefore has to consider the consequences of deciding to plant the wrong crop. We might quantify the loss in the following manner.

Under ideal weather conditions our farmer friend, call him Eldon, can expect a yield of 110 bushels of corn per acre (bu/acre). If the spring planting season is unusually wet and Eldon cannot plant until later in the spring, he will have to change to a lower yield, early harvest corn variety. Let us say that this other variety of corn will yield only 80 bu/acre. Unusually dry weather will adversely affect his yield, reducing his anticipated yield to, perhaps, 55 bu/acre.

Under ideal weather conditions Eldon expects a soybean yield of 45 bu/acre. If the summer brings cloudy, wet weather, his yield could theoretically be cut to 40 bu/acre. Dry weather will reduce Eldon's anticipated yield to, let's say, 30 bu/acre.

Alfalfa (hay) is measured in tons (T). Eldon can expect a 4 T/acre yield under ideal weather conditions. If, after cutting but before baling, an alfalfa field is rained on, the yield is not affected although quality will decrease. (Leaves of the alfalfa plants would be washed off

**TABLE 3    Gain for Farming Example**

| Weather | Corn | Decision to Plant Soybeans | Alfalfa |
|---------|------|----------------------------|---------|
| Wet | 80 bu | 40 bu | 4 T |
| Ideal | 110 bu | 45 bu | 4 T |
| Dry | 55 bu | 30 bu | 3 T |

the plants, thereby reducing quality.) Let us say that dry weather will reduce alfalfa yield to 3 T/acre.

Thus, our friend Eldon might quantify his **gain** for making different planting decisions as shown in table 3. (We can speak of loss or gain incurred by the decisions we make.)

The third component to a decision problem is information. Eldon has at his disposal the market price usually offered on these three crops. Since local weather conditions of concern to Eldon are not likely to affect corn prices, let us say that he might expect to receive $1.80 per bushel for his corn crop. (Widespread bad weather across Illinois or Iowa, for example, could affect corn prices, but local weather conditions aren't likely to have much of an effect.) Soybeans can be expected to bring $4/bu for all (local) weather conditions. On the other hand alfalfa quality might be affected by local weather conditions. Let us say that the hay will sell at $40/T under ideal or dry weather but only for $35/T if the weather is wet.

Eldon has been farming for a long time and has some very strong (prior) feelings about the chances of either wet, ideal, or dry weather for the planting season under consideration. Let us say that he will give 7 to 3 odds in favor of ideal weather — that is, he considers the probability of ideal weather to be 70%. Likewise, we will say that Eldon considers the chances of dry weather to be 10% and wet weather to be 20%. (Observe that quantification of these feelings by Eldon can be quite difficult; and, in practice, he may only indirectly use these feelings in reaching a decision.)

Now we are ready to decide what to plant. If Eldon plants corn, he can expect a cash receipt of

$80 \times \$1.80 = \$144$/acre if the weather is wet,

$110 \times \$1.80 = \$198$/acre under ideal weather conditions, and

$55 \times \$1.80 = \$99$/acre if dry weather occurs.

Since he feels that the weather will be wet, ideal, or dry with probabilities .2, .7, and .1, respectively, he can expect to receive

$$.2 \times \$144 + .7 \times \$198 + .1 \times \$99 = \$177.30/\text{acre if he plants corn.}$$

Similarly, he will receive

$$.2 \times 40 \times \$4 + .7 \times 45 \times \$4 + .1 \times 30 \times \$4 = \$170/\text{acre if he plants soybeans,}$$

and

$$.2 \times 4 \times \$35 + .7 \times 4 \times \$40 + .1 \times 3 \times \$40 = \$152/\text{acre if he plants alfalfa.}$$

Eldon will go with a corn crop. Good luck!

## Summary

We have examined the concept of probability inherent in statistical inference. Given that an entire population cannot be examined, we investigate a part of, or a sample from, the population. Inferring from the sample to the population involves us with uncertainty. Probability gives us a measure of this uncertainty.

We have discussed relative frequency probability as well as the probability associated with data having a (near) normal distribution. The relative frequency probability of an event is the number of occurrences of the event divided by the number of times the experiment is run. We stipulated that the experiment be a random experiment and that it be run a "large number of times."

Our discussion of normal probabilities centered around the random selection of a number from a normal data set. The probability of major interest is the probability that this number be within two standard deviations of the mean of the normal data. This probability is .95. In the next chapter we shall see how the normal probabilities discussed here fit into inferential statistics. Inference to characteristics of a dichotomous population, which was described in this chapter, will also be approximated using normal probabilities!

In chapter 5 we will first look at ways in which a sample can be selected from a population. The latter sections of the chapter will tie the normal probabilities summarized in table 1 of Appendix B to the uncertainty of statistical inference. In chapter 6 we will see how normal probability is related to the inferential processes of estimation and hypothesis testing.

## No Comment

■ **Congressmen Panic At Thought of Making Up Their Own Minds "Polling Encounters Public Resistance: Decision-Making Process Is Threatened"** — *New York Times Headline.*

"When the above story broke this week, revealing that the American people are tired of being surveyed, politicians in Washington and elsewhere were thrown into a sudden panic.

" 'This could precipitate a national crisis if we in Congress are forced to make up our own minds,' said one senator. 'I don't think anyone wants to see that. Or so my pollsters tell me.'

"A representative who contends he hasn't had an opinion he could call his own for 37 years, fears a lack of opinion polls could force him into early retirement. 'I make all of my hardest decisions based on the polls. Doesn't everybody?' "

(G. Wachman, "Without Polls, . . .",
New York News, October, 1975)[8]

# Exercises

1.  Determine the mean and standard deviation of the following data sets:
    a.  0, 0, 1, 1, 0, 0, 0, 0, 0, 0
    b.  1, 1, 0, 0, 1, 1, 1, 1, 0, 0, 0, 0
    c.  0, 0, 0, 1, 1, 1, 0, 0, 0, 0, 0, 0

2.  Suppose that a company manufactures steel rods. The rods are supposed to be 4 inches in length but the exact length of each rod will vary. Some rods will be slightly shorter than 4 inches; others will be longer than 4 inches. Suppose that the actual length of the rods produced by a company form a normal set of data with mean 4 inches and standard deviation .01 inch. A rod is considered "defective" if its length is less than 3.98 inches or greater than 4.02 inches. What is the probability that a rod manufactured by this company will be "defective"?

3.  In the previous problem one can think of the population of rods manufactured by the company as a dichotomous population. Here, a 1 represents a "defective" rod, and a 0 represents an "acceptable" rod. What are the mean and standard deviation of the 0s and 1s in this dichotomous population?

4.  Suppose that the seats on an airplane are 18 inches wide. Also, suppose that the widths of seats required so that passengers are comfortable form a normal set of data with mean 15 inches and standard deviation 1 inch. What is the chance that a passenger will find the seat too narrow — that is, what is the probability that a passenger will require a seat wider than 18 inches?

5. Suppose that an IQ club requires that members have an IQ greater than 145. If IQ scores are assumed to be normal with mean 100 and standard deviation 15, what are the chances that a person selected at random would qualify to join the club — that is, what proportion of the population are geniuses?

6. Recall our interpretation of the standard normal data graphed in figure 4(b). A value of $-3$ was three standard deviations below the mean. Can data from, let's say, a normal data set with mean 50 and standard deviation 10 be thought of in this way? What datum in the latter case corresponds to the $-3$ in the standard normal case? To $+2$?

7. The probability that at least 2 of 30 people have the same birthday is about 7 in 10. If a teacher "bet" a class of 30 that at least 2 students have a common birthday, what odds should be given to make the bet "fair"?

8. (Optional) An executive of a company must decide among three proposals for introducing a product, let's say a detergent, to an area. The three proposals involve: a television ad campaign; newspaper/radio ads; or putting free samples in as many households as possible.

   An ad campaign is considered successful if people in the area buy enough of the detergent to cover production, distribution, and advertising costs and provide a low profit. If the ad campaign is not successful, the company will hypothetically lose $10,000 for the television blitz, $6,000 for the newspaper/radio campaign, and $15,000 if they decide to put free samples in the homes of the area. The company will profit (after costs) by $3,000, $5,000, and $4,000, respectively, if the ad campaigns are successful.

   The executive feels that the chances of success for the three campaigns are .5, .3, and .2, respectively.

   Which ad campaign should the executive choose?

9. Find the following probabilities or values:
   a. The probability that a chi-square (8) value exceeds 20.1
   b. The probability that a chi-square (15) value is less than 25
   c. The value that is exceeded by 5% of chi-square (19) values

# For
# Discussion

1. Recall our discussion of IQ scores in chapter 2. Given that the results of a particular IQ test are normal with mean 100 and

standard deviation 15, how would your interpretation of such scores change from your previous thoughts? How would your interpretation of IQ scores as discussed at the end of chapter 2 not change?

2. A local newspaper reported that the probability of rain was zero on the day that 1.42 inches of rain fell (with damaging hail) on the area. How can this be?

3. How might we decide whether a collection of data is normal in distribution or nearly so?

4. Consider the decision of a television executive to cancel or retain a television program for another viewing season. Discuss the three components of, or influences on, this decision: prior feelings, fear of an incorrect decision, and statistical (and nonstatistical) information available to the executive.

5. A public television (statewide) network needs to make many important programming decisions. What type of information would be needed in order that good decisions be made? How might this information be obtained? Any problems involved with attempts to obtain this information?

6. A lawyer asks his client to, let's say, get a haircut, a shave, and dress up in a suit and tie. Which component of the decision-making process of the jurors is the lawyer trying to influence? How else might a lawyer try to get jurors to render a favorable decision?

# Further Readings

If our brief look into the history of probability has aroused your interest, consider F. N. David's *Games, Gods* and *Gambling* and L. E. Maistrov's *Probability Theory: A Historical Sketch,* both of which are cited in the Notes section of this chapter. Maistrov is, to my knowledge, the only author who has written about the development of probability as it relates to *both* gambling and the insurance business.

An interesting collection of essays on probability, game theory, decision theory, and cybernetics is available. This collection is *Mathematical Thinking in Behavioral Sciences,* readings from *Scientific American,* ed. D. M. Messick (San Francisco: Freeman, 1968).

If you'd like to read further into the subject of interpreting probability, consider L. J. Savage's *The Foundation of Statistics,* which we mention in the chapter Notes and H. E. Kyburg and H. E. Smokler's *Studies in Subjective Probability* (New York: Wiley, 1964).

# Notes

1. L. E. Maistrov, *Probability Theory: A Historical Sketch* (New York: Academic Press, 1974) pp. 4–5.
2. F. N. David, *Games, Gods and Gambling* (London: Charles Griffin & Company, 1962), p. 5.
3. *Ibid,* p. 7.
4. H. Zeisel and H. Kalver, Jr., "Parking Tickets and Missing Women," *Statistics: A Guide to the Unknown,* ed., J. M. Tanur (San Francisco: Holden-Day, 1972), pp. 102–111.
5. L. J. Savage, *The Foundations of Statistics,* (New York: Dover, 1973), p. 2.
6. R. C. Miller, "The Probability of Rain," *Statistics: A Guide to the Unknown,"* ed., J. M. Tanur, (San Francisco: Holden-Day, 1972) pp. 372–384.
7. Karl Pearson, "Historical Note on the Origin of the Normal Curve of Errors," *Biometrika* 16 (1924): 402.
8. Copyright 1975 New York News Inc. Reprinted by permission.

# The Sample Survey

5

THERE ARE TIMES WHEN EVEN well-known polls do not accurately predict what will happen in an election. For example, Harry Truman beat Thomas Dewey for the presidency in 1948 contrary to what the polls had predicted.

One of the reasons for the inaccuracy of polls is that the sample is not properly selected. Specifically, the sample is not selected in a way that affords a reliable measure of the uncertainty of statistical inference. We will, in this chapter, look at random sampling methods that make possible a reasonably reliable measure of the uncertainty of statistical inference.

In order that we might be able to measure the uncertainty of statistical inference we will propose a normal probability model and discuss when this model is appropriate. Most pollsters incorrectly assume that the model we propose is applicable to their sampling methods. We will discuss the sampling methods that are appropriate for this model so that you can detect misuse of the model.

As we proceed, recall that in chapter 2 we discussed background influences on sample survey statistics. These influences were the source, the population, the response rate, the method of contact, the timing of the survey, and the wording of questions. Two other factors we mentioned as having an important influence on our interpretation of survey statistics were the sample size and sampling procedure. All 8 of these factors should influence our interpretation of survey statistics. Part of proper interpretation of survey statistics will be to determine if the normal probability model usually assumed is indeed appropriate. Any one of these 8 factors might lead us to feel that the normal probability model is not appropriate, leaving us without a measure of the uncertainty of statistical inference.

B. C. by permission of John Hart and Field Enterprises, Inc.

## Scientific Polling

In chapter 3 we discussed census taking. There we noted that the word "census" is a misnomer because a complete enumeration of a population is impossible. (Recall table 2 in chapter 3, which summarized the extent to which the populace is not covered by the Census of Population and Housing.) When we talk of a census, we are therefore referring to an attempt at a complete enumeration of a population. The degree of completeness of the enumeration will tell us how credible the census is.

Recall also our discussions in chapter 1 and chapter 4 of how we are often unable or unwilling to attempt a complete enumeration of a population. Instead, we concentrate on looking at a part of, or sample from, the population. Inferential statistics are then numbers obtained from a sample that are used to infer to corresponding population numbers called *parameters*.

Before 1932 sampling a population was an informal process. If, for example, a politician wanted information from his constituents, he talked to a few of them. There was no rhyme or reason for the selection.

Mrs. Alex Miller may have been the first candidate for public office to have had the benefit of a preelection poll or information from a specially selected sample of the electorate.[1] Mrs. Miller was running for secretary of state of Iowa in 1932. A young man persuaded her to let him try a new sampling technique to survey Iowa. This new technique had been developed as part of the pollster's Ph.D. dissertation at the University of Iowa. The poll accurately predicted Mrs. Miller's win as Iowa's first woman secretary of state and pinpointed an impending Democratic sweep of Republican Iowa.

The young pollster, Mrs. Miller's son-in-law, was George Gallup. Dr. Gallup's university degree was in journalism. In an attempt to find a more accurate means of determining newspaper readership, he developed a sampling technique that is extensively used in opinion polling today.

When we speak of *scientific* sampling we mean sampling in a way that affords us an opportunity of measuring the uncertainty of statistical inference. Without scientific polling, survey results could well be as cartoonist Hart indicates in the cartoon at the beginning of this section. Before discussing such sampling procedures, let us first discuss how polls can be wrong.

## Faulty Preelection Polls

One of the major sources of statistical information is the sample survey. During elections many polls are taken. Some of these polls are taken by the candidates themselves. In this instance we get very little information on the results of the survey and even less of the information required to judge the credibility of the survey. (Recall our discussion in chapter 2 of the information we require in order to evaluate a survey properly.) At best we get only the results favorable to the candidate. Other preelection polls are common. Newspapers, newsmagazines, and syndicated polling organizations such as the Gallup and Harris groups get into the survey business during election time.

Preelection polling is unique to inferential statistics. A sample of adults who are registered to vote and intend to vote form the basis for inference to the population of adults who will vote on election day. This type of inference is of special interest since the unknown parameter of interest, the proportion of votes cast for a candidate in an election, will become known. These polls are then judged by the public, who knows exactly how accurate each survey was.

There are many examples of inaccurate preelection polls, some of which you have probably heard about. Examples include the 1936 election of Franklin D. Roosevelt over Alfred E. Landon and the 1948 election of Truman over Dewey. In both cases the preelection polls did not correctly predict the winner. The United States does not, however, have a premium on inaccurate preelection polling. In Canada Pierre Elliot Trudeau was elected prime minister in 1974 contrary to the prediction of the polls. Edward Heath's 1970 election over Harold Wilson in Great Britain was not unlike the 1948 United States polling disaster.

We will consider the polling for the 1936 and 1948 presidential elections in some detail. These examples should help us to evaluate sample surveys generally.

### 1936 Preelection Polls

For 20 years prior to 1936 the *Literary Digest* was quite successful in predicting the winner in presidential elections. Up to 20 million cards were mailed and as many as 3 million returns indicated the respondents' preferences in an upcoming election. Based on these large sample sizes of about 3 million, the *Digest* predicted a winner.

In a 12 July 1936 newspaper article George Gallup predicted 56% of the vote for Roosevelt over Landon, indicating that the *Digest* would be wrong in its upcoming polling attempt! Wrote the *Digest*, "Never before has anyone foretold what our poll was going to show even before it started. Our fine statistical friend [George Gallup] should be advised that the *Digest* would carry on with those old fashioned methods that have produced correct forecasts exactly one hundred percent of the time."[2]

The 1936 *Digest* poll was, as Gallup had warned, a disaster. The *Digest* predicted that FDR would receive 49.9% of the popular vote and

161 out of 531 electoral votes. He actually received 60.2% of the popular vote and 523 votes of the electoral college!

The FDR landslide ended the *Digest*'s preelection polling and contributed to its demise a few years later. There are three major reasons why this preelection poll was such a disaster. First, it was a voluntary response survey. People chose to be in the sample by mailing back the postcard sent them by the *Digest*. In the 1936 poll only 23% (2.3 million of 10 million) of those receiving cards returned a "ballot." Hence, 77% of the sample did not respond.

Second, the list of names from which the 10 million cards were sent was taken mainly from subscription lists of magazines, from telephone directories, and from lists of automobile owners. Such lists were biased toward people of higher income. This is especially true since we are talking of depression times. Since the more educated are more inclined to return a questionnaire, the sample was also biased toward this group. Hence, even though the sample was of substantial size, nearly 2.3 million for the 1936 poll, it was not representative of the population. The sample reflected the opinions of a more well-to-do, more highly educated segment of the population and was not representative of the opinions of the voter population as a whole. Gallup saw this bias developing in the *Digest*'s straw vote.

It should also be noted that this example shows that the sampling procedure is more important than the size of the sample. This can be seen by comparing the *Digest*'s sample of 2.3 million to today's national polls of about 1,200 people that are often more accurate than the *Digest* poll, mainly because of their sampling procedure.

The third reason that the *Digest* poll failed to predict the winner was that voting trends were not properly analyzed. It takes much time for 10 million cards to be mailed and returned. During this period voter preference shifted from Landon to Roosevelt. Had the *Digest* properly evaluated this change in voter preference it would not have been so far off. All the returns over this extended period were lumped together. This total picture showed Landon ahead, while masking Roosevelt's increasing popularity.[3]

## 1948 Preelection Polls

Another preelection polling disaster occurred in 1948. From the time of the last *Literary Digest* poll in 1936 until the Truman-Dewey campaign of 1948, opinion polls experienced much success and public support in the preelection polling arena. But in 1948 the major polling organizations predicted a Dewey victory and suffered a severe setback in public acceptance when Truman was elected. We see from table 1 that three major polling organizations erred from 5% to nearly 12% in predicting the vote for Truman.

**TABLE 1**    Percentages of Presidential Votes in 1948

|  | Dewey | Truman | Thurmond | Wallace | Other |
|---|---|---|---|---|---|
| National Vote | 45.1 | 49.5 | 2.5 | 2.4 | .6 |
| Preelection Polls |  |  |  |  |  |
| Gallup | 49.5 | 44.5 | 2.0 | 4.0 | — |
| Crossley | 49.9 | 44.8 | 1.6 | 3.3 | .4 |
| Roper | 52.2 | 37.1 | 5.2 | 4.3 | 1.2 |

Note: The pollsters used a quota sampling procedure in these 1948 preelection polls.

Source F. Mosteller, *The Pre-election Polls of 1948, Report to the Committee on Analysis of Pre-election Polls and Forecasts,* (Washington, D.C.: Social Science Research Council, Bulletin 60, 1949): 17.

The first of the two major reasons for the failure of these polls was, as in the 1936 example, the fact that the sample was biased toward the educated. Table 2 illustrates this point. We must be careful when interpreting the information in table 2. Our interest is with people who vote. The education of voters could well be different from the education of the general adult population, which is given in table 2. Gallup adjusted for the education difference (a technique discussed later in this chapter), but Elmo Roper did not adjust.[4] Roper contended that his sample reflected the education of people who voted. The wide variation between sample and population percentages does, however, suggest that a problem existed in this area.

The second and most important reason for the poor showing of the preelection polls of 1948 was "the failure of the polls to detect shifts in voting intentions during the later stages of the campaign."[5] We recall that the population of interest in a preelection poll does not exist at the time that the poll is taken. The population of interest consists of people who vote on election day. Therefore, the closer to election day that a sample is taken, the more likely the sample is to be representative of this nonexistent population. Roper declared, "As of this September 9, my whole inclination is to predict the election of Thomas E.

**TABLE 2**    Sample Percentages

| Education (Last School Attended | Population Estimate | Gallup (14 October Sample) | Roper (25 October Sample) |
|---|---|---|---|
| Grade School or Less | 43.5 | 35.3 | 27.5 |
| High School | 43.4 | 46.8 | 48.8 |
| College | 13.0 | 17.9 | 23.7 |

Source: Mosteller, *The Pre-election Polls of 1948,* p. 105.

Dewey by a heavy margin and devote my time and efforts to other things."[6] Gallup and Archibald Crossley conducted their last surveys two weeks prior to the election. Yet, 1 in 7 voters made their decision in the last two weeks, with 3 out of 4 deciding to vote for Truman. Hence, Truman gained the support necessary for victory between the time of the last poll and the actual election.

Since 1948, preelection polls try to detect voter shifts near election day. You will notice that the last poll before an election is now made public the day before the election or even on election day itself, with polling conducted only a few days before the election. This last poll might be a telephone poll that can be completed in one day, let's say, the Sunday before election day. It is thought that in this way the population sampled will be more like the population of interest (see Discussion 9 of this chapter).

Another point of interest, which is closely tied to our discussion of late voter-shifts, concerns the pollsters' preelection handling of the undecided voter in 1948. There are people who have not yet decided how they will vote when the preelection poll is taken. For the 1948 campaign, people who were undecided were allocated to the candidates on a proportional basis, that is, they were eliminated from the sample altogether.[7] This undecided group was of substantial size prior to the 1948 election — up to 15% (the 1 in 7 previously mentioned). Today the "undecided" are carefully watched.

In retrospect, the pollsters of the 1948 election had backed themselves into a corner. Their press releases indicated that they were so positive that Dewey would win that they did not leave themselves an out. Assimilators of statistical information should take note. There is no certainty in statistical inference; there is **uncertainty**.

As an aside, note that not all pollsters were wrong in 1948. Consider the following:

> One correct poll: Staley Milling Co., Kansas City . . . is blushing this one off — not because it was wrong, but because it was right.
>
> Just before the election the company held an informal poll in which farmers cast their votes by calling for a chickenfeed sack with a donkey or an elephant on it. After three weeks, the results stood: 54% for Truman, 46% for Dewey. Unable to believe what it saw, Staley called the thing off. Quipped one executive at the time (with more foresight than he realized), "If pullets were ballots, President Truman would be a shoo-in for re-election."[8]

We see how important it is that the sample be representative of the population. Accurate surveys can be conducted with samples of size 1,200, which indicates that the size of the sample is not as important as the method of selection. (Recall that the **Literary Digest** had nearly 2.3 million respondents in 1936.) Assuming, then, that a sample is representative of a population, we must be sure that the inference is made only to that population.

## Simple Random Sampling

We will now begin our discussions of scientific sampling procedures. By scientific we mean sampling procedures that make it possible to measure the uncertainty of statistical inference. After discussing different types of sampling procedures, we will discuss the normal probability model. Our discussion of the applicability of this model will be conducted in this and the succeeding chapter.

> A sample drawn from a population is a *simple random sample* if all samples of the same size have an equal chance of being the sample.

The term "random," when used to refer to sampling a population, indicates that an outside chance mechanism is used in the selection of elements from a population. We will discuss such a mechanism later in this section.

Before proceeding further we need to distinguish between observational units and sampling units.

> *Observational units* are the smallest elements on which an observation is made in a sample.
>
> *Sampling units* are the smallest elements used in the selection process of a sample survey.

For example, suppose we select households but talk to everyone in the household. The household member interviewed is the observational unit, and the household is the sampling unit. If we asked questions concerning, let's say, the number of television sets in the household, the household is both the observational unit and the sampling unit.

Let us look at the selection of a simple random sample from a population. Suppose, for purposes of illustration, that we want to select a simple random sample of size 2 from a population of 6 elements. We will denote the 6 elements in this population by the letters *A, B, C, D,*

*E*, and *F*. There are 15 distinct ways that 2 objects can be selected from this collection of 6 elements. The samples are:

$$(A,B), (A,C), (A,D), (A,E), (A,F), (B,C), (B,D), (B,E),$$
$$(B,F), (C,D), (C,E), (C,F), (D,E), (D,F), (E,F)$$

(Observe that the sample $(A,E)$ is considered the same as the sample $(E,A)$, that is, the order in which elements are selected is not important.) A simple random sample of size 2 taken from a population of 6 elements requires that each of the 15 samples (all possible samples of size 2 that can be selected from 6 elements) have an equal chance, 1 in 15, of being the sample. For example, $(A,E)$ is just likely to be the sample as is $(B,F)$.

## Sampling Without Replacement

Note that we are selecting *without replacement* — that is, after an element (a letter) is selected, it cannot be selected again since it is not replaced into the population before the next element is selected. When selecting without replacement, a simple random sample can be obtained by assuring that at each selection all remaining elements have the same chance of being selected into the sample. In our example this means that when the first letter is selected, each of the 6 letters has an equal (1 in 6) chance of being selected. When the second letter is selected from the remaining 5 letters, each of the 5 has an equal (1 in 5) chance of being selected. There are many random sampling techniques in which each element in the population has the same chance (overall) of being in the sample. Simple random sampling is one such selection procedure.

This process can be extended to any size simple random sample that is to be selected from any size population. Since the size of the sample and population can be quite large, we look for a method of extracting a simple random sample of any size from any population. In order that a sample may be drawn from a population, a list of the elements in the population is needed.

---

A *frame* is a complete description of a population.

---

This description could be a list of population elements. Using such a list a simple random sample can be selected.

Given a list of elements in a population, we can extract a simple random sample of elements by first numbering the elements in the population. We then select a simple random sample of numbers, which gives a simple random sample of elements — those elements corresponding to the selected numbers.

### Random Digit Table

Let us consider the simple example of the population of 6 elements, **A, B, C, D, E,** and **F.** Number these 6 elements successively from 1 to 6, that is,

$$A \longleftrightarrow 1$$
$$B \longleftrightarrow 2$$
$$C \longleftrightarrow 3$$
$$D \longleftrightarrow 4$$
$$E \longleftrightarrow 5$$
$$F \longleftrightarrow 6$$

We then select a simple random sample of size 2 from the numbers 1, 2, 3, 4, 5, and 6. This can be accomplished by using a **table of random digits**. Table 3 of Appendix B is a listing of 2,500 random digits, 50 rows of 50 columns each. A section of this table is given in table 3 in this section.

Each digit in table 3 has a specific column and row designation. Each group of 5 columns forms a block in the table. This is done for easier reading. We start by reading numbers from the table from a random starting point. It suffices to place a finger (randomly) on the table, noting the nearest four-digit number: The first two digits represent the row, and the second two digits represent the column where we will start. Suppose we get the number 3335; we will begin in the 33rd row, the 35th column. Note that since we have only 50 rows and 50 columns, the first and third digits of this four-digit number must be

TABLE 3    Random Digits

Column

35

| | |
|---|---|
| 67384 | 95693 |
| 06845 | 82326 |
| 43597 | 07110 |
| 60709 | 29931 |
| 7319① | 10822 |
| 74409 | 43336 |
| 72166 | 37839 |
| 75625 | 48777 |
| 34191 | 40169 |
| 37149 | 92805 |

Row 33

less than 5 in order to be useful. Therefore, if the first or third digits are more than 5, subtract 50, that is, 85 represents row 35, 35 = 85 − 50.

Table 3 shown in this chapter gives us the 33rd row and the 35th column of table 3 of Appendix B. The digit in the 35th column and 33rd row is 1. (See the circled 1 in table 3 of this chapter.) This is the first *number* in our simple random sample. Reading down, we then read the (underlined) value of 9, which is ignored since our range of interest consists of the numbers from 1 to 6. Continuing down the 35th column, we see an (underlined) 6, which is in the range. Our simple random sample of size 2 from the numbers 1, 2, 3, 4, 5, and 6 is then (1,6). The corresponding simple random sample of elements would be the sample (*A, F*). (Had we encountered a second 1, a number already selected, we would have ignored it. Recall that we are sampling without replacement so once a number [element] is selected it cannot be selected a second time.)

Note that if the range of numbers from which we were to select a simple random sample were greater, let's say, 000 to 347, we would use *3* columns from table 3 of Appendix B, rather than 1 column. We would think of these numbers as 000, 001, 002, 003, etc. From the random starting point of the 33rd row and 35th column we would read three-digit numbers from the block containing column 35. Note that these three-digit numbers would begin with the number 191 (the first number in our sample), then, continuing down columns 33–35 to 409 (not in range, so ignored), 166 (second number in sample), 625 (not in range), 191 (ignored as already in sample), etc.

Given that the digits in table 3 of Appendix B are random, observe that the sample of numbers selected by the procedure above is indeed a simple random sample — as each member of the sample is selected, all unselected numbers have an equal chance of being the next number in the sample.

## Other Sampling Procedures

We will now discuss 4 other sampling techniques: prestratification, poststratification, cluster sampling, and systematic sampling. Prestratification and poststratification are techniques that can help to ensure that a sample is representative of a population; cluster and systematic sampling are money-saving techniques.

### *Stratification and Prestratification*

When selecting simple random samples, certain groups within the population may be overrepresented and other groups underrepresented. For example, for a preelection survey a greater proportion of females may be in the sample than is desired. We would like the proportion of women sampled to be near the proportion of women expected to vote. This proportion would be near the proportion of women who voted in the last election. By chance, however, more women than desired may end up in a simple random sample. A more desirable sampling approach would be to take two separate simple random samples, one of females and one of males. This type of sampling procedure is a (pre)*stratified random sample.*

To help ensure that a sample is representative of the population of interest, a population may be divided into parts called *strata*. A sample is then taken from each *stratum*, ensuring that each stratum is represented in the overall sample. Stratification is the process of dividing a population or a sample into strata. *Stratification* is the division of a collection of elements into parts called strata. *Prestratification* refers to stratification of a population before a sample is selected. Each stratum is then sampled. The division can be based on geographic, economic, and/or social characteristics of the population. The type of stratification depends on the survey. If it is desirable that different economic groups be represented properly in a sample, a population would be stratified on economic grounds. People of roughly similar economic status would be put in the same stratum. (See chapter 8 for a discussion of survey design.)

An example of economic stratification is *census tracting*. Local committees from certain Standard Metropolitan Statistical Areas (SMSAs) set tract boundaries under the direction of the Bureau of the Census. Recall from chapter 2 that a SMSA is a collection of counties that touch a county containing a central (large) city. These tracts have a mean size of about 4,000 people and are originally laid out with the intention of achieving within the tracts some uniformity of population characteristics, economic status, and living conditions. Generally, census tracts should:

1. Contain between 2,500 and 8,000 inhabitants

2. Follow permanent and easily recognizable boundaries

3. Agree with county boundaries (towns and townships in New England states)

4. Contain people of similar economic status and housing as much as possible (expensive housing and slum housing should not both be contained in the same tract since averages would not reflect the condition of either group)

5. Have compact boundaries (not such configurations as panhandles, *L*s, dumbbells, etc.)

This type of stratification has been completed for a number of SMSAs. Using these tracts, it is possible to ensure that a sample properly reflects race or nationality characteristics as well as the social, economic, and housing standards of the inhabitants of the SMSAs.

Many samples are based on *geographic stratification*. As you might guess, the division of a population according to place of residence is easier to accomplish than stratification by socioeconomic characteristics. Besides census tracts, which are geographic as well as socioeconomic groupings, geographic stratification can be found in surveys run by national survey groups. For example, the Gallup organization ran a survey for the Zenith Corporation using geographic strata.

The United States was divided into these four size-of-community strata: (a) cities of population 1,000,000 or over, (b) 250,000 to 999,999, (c) 50,000 to 249,000, (d) all other population.

Within each of these strata, the population was further stratified by seven regions: New England, Middle Atlantic, East Central, West Central, South, Mountain, and Pacific. . . .

Population in the fourth stratum was arrayed by counties, with each county's population classified into these four substrata: urban fringe around cities in the first three strata, cities 2,500 to 49,000, rural towns and villages, and rural open country areas.

From this array a national sample of locations was drawn with the probability of a location's selection proportional to its size in the 1970 census.[9]

Given that a total sample size has been decided upon and that a prestratified random sample is to be selected, we still have to decide what part of the total sample will come from each stratum.

---

*Sample allocation* is a determination of the part of an overall sample that is to be taken from each stratum for a prestratified sample.

---

The number of sampling units drawn from each stratum may depend on the variability or spread of the numbers in the stratum, the cost of interviews, and/or the size of each stratum. If numbers in a stratum are relatively homogeneous (do not vary much in magnitude), a relatively small number of sampling units are selected from the stratum. If, on the other hand, the numbers in a stratum are quite spread out, a large number of sampling units may be required to get an accurate picture of the numbers in the entire stratum.

The cost of interviewing within a stratum may also affect the number of units selected from each stratum. More units are selected from strata for which the cost of interviewing is relatively small, with fewer sampling units being selected from strata for which interviewing is more expensive. Such an allocation of units to the strata is used when survey cost is an important consideration and when these costs vary greatly between strata.

Let us look at an example so we might have a better sense of what is involved. A railroad proposed sampling certain freight shipments to determine its revenue during a specified time interval and on a particular section of track.[10] A waybill, a document accompanying a shipment of freight, gives the required information for such a study. The population consisted of 23,000 waybills. The waybills in the entire population ranged from $2 to $200 in size.

The population was first divided into strata that reflected the size of the waybills. Table 4 shows this stratification. Also indicated in table 4 is the proportion of each stratum that is to be taken for the sample. Only 1% of the waybills in stratum 1 were selected since the small variation in amounts of these waybills indicates that a small sample provides adequate information on that stratum. In contrast, all of stratum 5 were included in the sample; more information is needed concerning stratum 5 because of the great variation in the amounts of the waybills in this stratum.

Think of this type of allocation by considering stratification for the purpose of auditing a bank. All accounts of over, let's say, $100,000 would be sampled (audited) since an error found in one such account could affect the entire audit. (See Discussion 1 of this chapter.)

Allocation *proportional to strata size* is often used to determine how many observations are to be taken from each of the strata. Proportional allocation means that if, for example, stratum 1 contains twice as many elements from the population as does stratum 2, twice as many sampling units are taken from stratum 1 as are taken from stratum 2. The Gallup survey run for Zenith Corporation, which we mentioned earlier, is a case in point. Proportional allocation was used to determine how many observations would be taken from the various

**TABLE 4    Stratification for Waybills Example**

| Stratum | Waybills Charges Between | Proportion to Be Sampled |
|---|---|---|
| 1 | $   .00 and $  5.00 | 1% |
| 2 | $  5.01 and $10.00 | 10% |
| 3 | $10.01 and $20.00 | 20% |
| 4 | $20.01 and $40.00 | 50% |
| 5 | $40.01 and over | 100% |

Source J. Neter, "How Accountants Save Money by Sampling," *Statistics: A Guide to the Unknown* ed. J. M. Tanur (San Francisco Holden-Day, 1972), pp. 203–211.

area strata. Notice that this type of allocation requires knowing the size of each stratum. Strata boundaries usually agree with census boundaries such as counties, census tracts, and blocks. Strata sizes are then available from census records. Up-to-date census records are therefore important.

The way that sampling units are selected within the strata will determine whether a total sample is a *quota sample* or *a stratified random sample*. If a simple random sample is selected within each stratum, the resulting (overall) sample is called a stratified random sample.

A quota sample makes no demand on the selection procedure used within the strata: elements are selected from each stratum by any nonrandom means. For example, suppose that 13% of a population has a college education. An interviewer is then asked to interview any 130 people with a college education for a sample of size 1,000 (13% of 1,000 is 130). Observe that people with a college education represent a stratum. For the waybills example, a quota sample would require that 10% of stratum 2 be selected by any method. If the sample selected from **each** stratum is a simple random sample, the total sample would be a stratified random sample.

## Poststratification

> *Poststratification* refers to division of a *sample* into parts or strata. (Recall that prestratification refers to division of the population into strata. The strata are then sampled.)

It is often not possible to stratify a population before the sample is selected. Poststratification is carried out only when prestratification is practically or economically impossible. For example, consider a survey on the viewing of public television. Whether people watch public television depends, to a large extent, on their income, their education, and the number of preschool children in their household. (See chapter 8 for a detailed discussion of this type of survey.) Households cannot be segregated according to these factors until after the sample is drawn. (Only the Census Bureau would have the necessary information for prestratification on these variables. This information is not available to pollsters.) After the sample is drawn, you would want to check and see if the sample is indeed representative with regard to these factors.

As an example, let us look at one of the closest senatorial races in history. In 1974 Louis Wyman received two more votes than John

Durkin for a senate seat from New Hampshire. (Durkin later won the seat in a special election when the United States Senate could not decide who "won" the first election.) A late poll concerning the regular election reported Durkin trailing by 17%.[11] Such a gap between candidates can have a stifling effect: contributions dry up and fieldwork comes to a near standstill.

The poll, however, was based on a sample containing 1/6 independents. New Hampshire has 1/3 independents! Had the survey results been weighted to reflect the proper level of independents, the results would have more accurately predicted this very close race. The fact that the sample had only 1/6 independents should have signaled that the sample may not have been representative of New Hampshire voters. Weighting the sample results by known levels of Democrats, Republicans, and independents in New Hampshire can be viewed as a futile attempt to salvage something from nothing. Poststratification should have signaled that the results of the survey should not "see the light of day."

Let us look at other uses of poststratification. Readers of a magazine were asked to return a questionnaire that contained some very probing questions on sexual behavior. Nearly 100,000 women returned the published questionnaire. The results of this survey were (implicitly, if not directly) extended to the entire United States population of women. The basis of this inference was not only the size of the sample (100,000) but also a poststratification of the 100,000 respondents. The 100,000 women who returned the questionnaire were categorized by age, marital status, number of children, religion, religious convictions, political attitudes, education, family income, working status, and size of community. When the percentage of respondents from a stratum was close to the census percentage, the survey results were presented as "representative." When the percentages differed, the reader was warned that the report was "least applicable." For example, results were least applicable to women over 50 years of age, to women who had only a gradeschool education, to women whose income was less than $9,000, to women who were not white, and to women who were unmarried. Poststratification was being used to try and convince readers of the magazine that the sample of 100,000 women was indeed a representative sample of all women.

Even if, let's say, 10% of the women returning a questionnaire are between the ages of 35 and 39 and about 10% of all women in the country fall within this age category, would these 10,000 respondents (10% of the total 100,000) be representative of all women aged 35 to 39? Remember, this is not a survey involving random selection; rather, this is a voluntary response survey. Are women aged 35 to 39 who return such a questionnaire "different" from women in this age category who chose not to return the questionnaire or who do not even subscribe to the magazine that published the questionnaire.

Be careful in this regard. A quota sample based on age is one whose percentages of sampling units in different age categories agree with population percentages. Selection within these categories is not governed by a random mechanism. Without some sort of random selection mechanism, the uncertainty of statistical inference cannot be measured and so the accuracy of inference to the population of interest cannot be properly judged.

Would the above survey be any more representative if responses in different categories were *weighted* by the size of these categories in the population? Suppose, for example, that 12% of women over the age of 40 answered yes to a question. We would weight (multiply) this percentage by the proportion of women in the country who are 40 years of age or older. (This weight would be about 38.1% or .381.) If such a weighting scheme were carried out, we could ensure that the *size* of each subgroup is properly reflected. But the problem remains that the subgroup samples would not necessarily be representative of all women in the population subgroups.

A radio survey gives us an example of a weighted sample and a final example of how poststratification is used and misused. Suppose that 11 age-sex categories are used: teens (ages 12–17); men (ages 18–24, 25–34, 35–49, 50–64, 65+); and women with the same 5 age categories as the men. Responses within a particular category are weighted by the proportion of the population in that category. For example, if 12% of the men between the ages of 35 and 39 listen to a particular radio station at a certain time, this percentage would be weighted by .102 if this is the proportion of men in a market area between the ages of 35 and 39. This technique of poststratification makes the reported percentages more representative of the number of people in an area in the different age-sex categories. With this system, if, let's say, a large percentage of women between the ages of 18 and 24 returned the diary, the percentages reported for these women are reduced by weighting, and groups under represented have their percentages increased in influence.

But the problem remains: Do the men between the ages of 35 and 39 who respond have opinions that are representative of the radio listening habits of all men in this age-sex category, regardless of whether the *number* of men in this category are properly represented? The response rate can be so low for such a survey that weighting cannot aid us with our inference to the entire population of interest. This becomes more like a quota sample of the entire population. For quota samples we do not have a measure of the uncertainty of inference.

Be suspicious of attempts to weight strata responses by "known" population weights. The problems with such a process are:

1. Accurate population weights may not exist, making weighting of questionable value.

2. The need to weight strata responses may be a signal that the sample was improperly drawn in the first place, and weighting is not likely to correct for an improperly obtained sample.

### Cluster and Systematic Sampling

Besides the simple random sample and stratification techniques just discussed, we will also consider cluster sampling and systematic sampling. These 4 types of sampling techniques are the ones most used in sample surveys.

> *Cluster sampling* involves the selection of clusters or groups of elements, rather than individual elements of the population. (Selection of clusters could be based on the size of the cluster or could be a simple random sample of clusters.)

For example, blocks of households may be selected rather than individual households. A simple random sample of areas of land (clusters) may first be taken using a map. We described the selection process in the last section. Clusters of elements, not individual elements, are numbered and then selected. Selection of area clusters is followed by the selection of households within the selected areas. In the types of surveys you are likely to hear or read about in the media, cluster selection is but one stage in a many-staged, or multistage, sampling design.

Cluster sampling may also be used in surveys that are not multistaged. For example, when sampling a corn field for the purpose of estimating yield, a sample of areas of the field or clusters of plants may be selected. Estimates of yield in these areas are then used to predict the total yield of a farm or field.

Systematic sampling is extensively used by the Census Bureau in conjunction with census taking as well as in the final stage of a multistage sample selected for, let's say, an opinion survey.

> A *systematic sample* requires that, from a random starting point, every so many elements in a population are selected.

For example, for a systematic (20%) sample of the population, every fifth household is selected. Hence, 1 in 5 or 20% of all households are in the sample and are asked questions not asked of *all* households in the complete population census. (Recall our discussion of this in chapter 3.)

After blocks have been selected in a multistage sampling procedure, an interviewer may be told to talk to every sixth household in a block (counting from a predetermined random starting point). This is a useful approach since a list of households in a block might not exist. An interviewer could select households without even knowing how many houses are on a particular block. If every sixth household is to be selected with a systematic sample, more households will be selected from larger blocks. We would be taking 1/6 of the households on each block or a fixed proportion from each block cluster. This is sometimes a desirable end product of systematic sampling, since it gives each household on a block in an urban area the same chance of being selected once the block had been selected.

Having looked at different random selection techniques, let us now see how this randomness can be converted into a probability measurement of the uncertainty of statistical inference.

## Distribution of the Sample Proportion

The uncertainty involved with using a sample statistic to infer to an unknown population parameter is measured by relative frequency probability. Recall from chapter 4 that relative frequency probability is involved with the execution of a random experiment. You can visualize such probability by considering what would happen if this experiment were conducted a large number of times. The experiment under consideration in statistical inference is the taking of a random sample — that is, some random mechanism is involved in the selection of a sample from the population of interest. In order that we can understand probability as it relates to statistical inference we must ask ourselves what happens when the random sampling procedure is repeated many times. Each time that a sample is drawn we will calculate a statistic from the sample. Our purpose is to describe this large collection of values of a statistic, one value for each of a large number of random samples. The distribution of these values of a statistic is called the *sampling distribution of the statistic*.

Let us look in some detail at a particular example. Consider a dichotomous population, that is, a collection of 1s and 0s. Recall that in a preelection poll a 1 could be "for Candidate A," or in a television-viewing survey a 1 could represent "a viewer." Also recall that the mean of all the 0s and 1s in the population turns out to be the proportion of 1s in the population. The sum of any collection of 0s and 1s equals the number of 1s. The mean of such a collection, which is this sum divided by the total number of 0s and 1s in the collection, equals the proportion of 1s.

$$\text{mean} = \frac{\text{sum of 0s and 1s}}{\text{number of 0s and 1s}} = \frac{\text{number of 1s}}{\text{number of 0s and 1s}}$$

$$= \text{proportion of 1s}$$

Because we are interested in the proportion of 1s in a dichotomous population we look at the sample proportion — that is, we would look at the proportion of 1s in a random sample, an inferential statistic, if we are interested in the proportion of 1s in the population, which is a parameter.

Suppose we are selecting simple random samples from a population. We note that the sample proportion may be a *different number* for each *different sample* that is selected. An important question needs to be considered: Can we describe the different values the proportion of 1s in a sample might take on for the different random samples that could be selected? Let us see.

### Probabilities for Sample Proportions

Suppose, for the sake of illustration, that one half of the elements in a large dichotomous population are 1s and, then of course, one half are 0s. If we take simple random samples of size 2 from this population, we could get either two 0s, one 0 and one 1, or two 1s. The *proportion of 1s* in these three distinct samples (the sample proportion) is 0, 1/2, and 1, respectively. We can find the probability that each of these different samples will occur. The *probability of each sample* will equal the probability that the sample proportion takes on the values 0, 1/2, and 1. These probabilities are 1/4, 1/2, and 1/4, respectively. Table 5 summarizes these results. (We will not go over the calculation of these probabilities.) A graph of the latter two columns of table 5, the

**TABLE 5    Samples of Size 2 from a Dichotomous Population**

| Sample | Sample Proportion | Probability* |
|---|---|---|
| Both 0s | 0 | ¼ |
| One 0 and one 1 | ½ | ½ |
| Both 1s | 1 | ¼ |

*Selection probabilities assume sampling *with* replacement but will be close to probabilities obtained if sampling were *without* replacement since our population is considered "large."

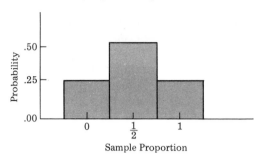

FIGURE 1
Distribution of the Sample
Proportion (for samples of size
2 taken from a dichotomous
population with 1/2 1s)

distribution of the sample proportion, is shown in figure 1. We use a histogram in figure 1 rather than a bar graph even though the sample proportion takes on only three values. As the sample size increases, the values of the sample proportion will tend to take on all values between 0 and 1; that is, become like a continuous set of data.

> A graph of the possible values of the proportion of 1s in a sample relative to the probability that this statistic will take on the values is called the **distribution of the sample proportion,** or the **sampling distribution of the proportion.**

We can think of the sampling process as generating a new collection of numbers (values that the statistic may take on) — in our example a collection of 0s, 1/2s, and 1s in ratio of 1:2:1. Each time that a random sample is selected the proportion of 1s in the sample is calculated. The sample will consist of two 0s, one 0 and one 1, or two 1s. The sample proportion will be 0, 1/2, or 1, respectively, and can be expressed in the manner shown in table 6. Using methods not to be presented here, it can be determined that the sample (0,1) is twice as likely to occur as the samples (0,0) and (1,1), which both occur equally often. Hence, the proportion 1/2 will occur twice as often as 0 or 1; 0 and 1 will occur equally often. Since probabilities must add to one, the probability that the sample proportion will be 1/2 is 1/2, and the probability that the sample proportion is 0 or 1 is 1/4 in both cases. A graph of the values that the sample proportion can take relative to the corresponding probabilities is the sampling distribution of the proportion. This graph is shown in figure 1.

For a sample of size 6, the collection of possible values of the sample proportion, an inferential statistic, is described in table 7.

Figure 2 represents the sampling distribution for the proportion

TABLE 6    The Sample and the Sample Proportion

| Sample | Proportion of 1s |
|--------|------------------|
| (0, 0) | 0 |
| (0, 1) | ½ |
| (1, 1) | 1 |

TABLE 7    Samples of Size 6 from a Dichotomous Population with 1/2 1s

| Sample | Sample Proportion | Probability |
|--------|-------------------|-------------|
| All 0s | 0 | $1/64$ |
| 5 0s and 1 1 | $1/6$ | $6/64$ |
| 4 0s and 2 1s | $2/6$ | $15/64$ |
| 3 0s and 3 1s | $3/6 = 1/2$ | $20/64$ |
| 2 0s and 4 1s | $4/6$ | $15/64$ |
| 1 0 and 5 1s | $5/6$ | $6/64$ |
| All 1s | 1 | $1/64$ |

when samples of size 6 are selected from a dichotomous population with 1/2 0s. Suppose that a large number of random samples of size 6 is selected and the proportion of 1s in each sample is calculated. The collection of values of the sample proportion would have a histogram that looks like figure 2.

Figure 3 is the graph of the distribution of the sample proportion for samples of size 30. A normal distribution is drawn on the graph of the distribution of the sample proportion. We see how closely the sampling distribution of the proportion can be approximated by a normal distribution when the sample size is 30.

If we continued to increase the size of the simple random samples, the graph of the values of the sample proportion would remain symmetric about 1/2 and would become closer and closer to a bell-shaped curve. Indeed, for large sample sizes, the distribution of the sample proportion becomes more and more normal! This amazing fact, first observed by de Moivre in 1733, allows us to describe the collection of values of the sample proportion for different samples without actually doing all the work summarized in table 5 and table 7. We could say that for large-size simple random samples the distribution of the sample proportion is about normal, with center 1/2 (the proportion of 1s is the original population).

Recall that the normal curve is completely specified when both the mean of the curve (1/2 in this case) the standard deviation are

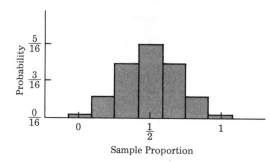

**FIGURE 2**
**Distribution of the Sample Proportion (for samples of size 6 taken from a dichotomous population with 1/2 1s)**

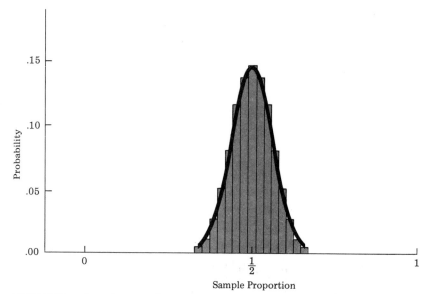

**FIGURE 3**   **Distribution of the Sample Proportion (for samples of size 30 taken from a dichotomous population with 1/2 1s)**

known. It can be shown that the standard deviation of the normal curve, which approximates the distribution of the sample proportion, is the standard deviation of the dichotomous population under investigation divided by the square root of the sample size.

$$
\begin{array}{c}
\text{standard deviation} \\
\text{of normal curve}
\end{array}
=
\dfrac{
\begin{array}{c}
\text{standard deviation} \\
\text{of dichotomous population}
\end{array}
}{
\begin{array}{c}
\text{square root} \\
\text{of the sample size}
\end{array}
}
$$

Recall in chapter 4 in the section "Two Distinct Groups: The Dichotomous Population" we said the standard deviation of a dichotomous population could be easily determined by referring to the figure in Appendix C. From Appendix C we note the standard deviation of a dichotomous population with 1/2 1s is 1/2. The standard deviation of the distribution of the sample proportion, for samples of size 100, is

$$\frac{.5}{10} = .05$$

The square root of 100 is 10. (Square roots are given in table 4 of Appendix B.)

For finite populations a correction factor for the standard deviation of the distribution of the sample proportion has not been given. This factor is the square root of the proportion of population elements *not* in the sample. The standard deviation of the normal curve is multiplied by this factor to correct for the fact that we are sampling from a finite population. If the proportion of elements not sampled exceeds .95, that is, less than 5% of the population was sampled, this correction factor (the *finite population correction*) is usually ignored. If the sample is more than 5% of the population, the finite population correction should not be ignored. We will not use this correction factor. (See Exercise 5.)

## When the Proportion of 1s is not 1/2

Our discussion thus far has been restricted to the case in which one half of the elements in a dichotomous population are 1s. As was seen, the distribution of the sample proportion is symmetric for all the sample sizes discussed. We would therefore expect the normal curve to be a good approximation of the distribution of the sample proportion for samples of moderate size.

If the proportion of 1s in the original population is, however, far from 1/2, the distribution of the sample proportion for small sample sizes is *not* close to normal. For example, suppose there are one-tenth 1s and nine-tenths 0s in a large dichotomous population. Table 8 summarizes results for selection of samples of size 6.

TABLE 8    Samples of Size 6 from a Dichotomous Population with 1/10 1s

| Sample | Sample Proportion | Probability |
|---|---|---|
| All 0s | 0 | .5314 |
| 5 0s and 1 1 | $1/6$ | .3543 |
| 4 0s and 2 1s | $2/6$ | .0985 |
| 3 0s and 3 1s | $3/6 = 1/2$ | .0145 |
| 2 0s and 4 1s | $4/6$ | .0012 |
| 1 0 and 5 1s | $5/6$ | .0001 |
| All 1s | 1 | .0000 |

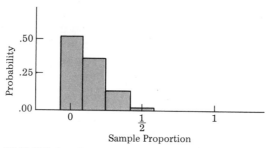

FIGURE 4    Distribution of the Sample Proportion (for samples of size 6 taken from a dichotomous population with 1/10 1s)

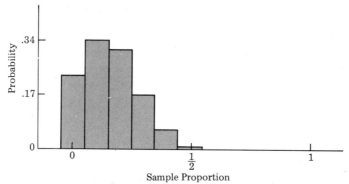

FIGURE 5    Distribution of the Sample Proportion (for samples of size 16 taken from a dichotomous population with 1/10 1s)

Figure 4 shows a graph of the distribution of the sample proportion for samples of size 6 taken from a dichotomous population with 1/10 1s. The graph in figure 4 is skewed to the right — not at all like the normal distribution. Even so, as our sample size gets larger (larger than in the case with 1/2 1s in the population), the distribution of the sample proportion becomes more and more like a normal curve.

We might be better able to visualize this idea by looking at the distribution of the sample proportion for samples of size 16 when 1/10 of the population values are 1s. Figure 5 illustrates such a distribution. From figure 5 we observe that the mode of the distribution of the sample proportion has shifted to 1/10 (toward the center of the graph). Indeed, if we were to increase the sample size even more, 1/10 will become the center of a *symmetric* distribution. The phenomenon can be visualized by looking at the sampling distribution of the proportion for samples of size 30 drawn from a population with 1/10 1s. We see in figure 6 that the sampling distribution of the proportion is indeed close to

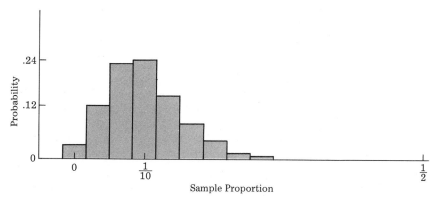

**FIGURE 6**    **Distribution of the Sample Proportion (for samples of size 30 taken from a dichotomous population with 1/10 1s)**

symmetric for samples of size thirty. We can, in fact, make the general statement:

> When simple random samples are drawn from a large dichotomous population in which **P** represents the proportion of 1s, the distribution of the sample proportion is **about** normal; the mean of this approximately normal distribution is **P**; and the standard deviation is the standard deviation of the population (given in Appendix C) divided by the square root of the sample size. The distribution of the sample proportion is closer to normal for larger size samples.

It is very difficult to set a value for the sample size required for a "good" normal approximation of the distribution of the sample proportion. If $P = 1/2$, a sample of size 10 to 20 would be large enough. Should the proportion of 1s come close to .1, a sample size of around 100 might be needed for an adequate normal approximation. (For even smaller **P**, other approximations besides the normal are recommended.[12] We'll assume that sample sizes are large enough for the surveys we are likely to hear or read about in the media, (see Exercise 7.)

## Some Examples

In our first example $P = 1/2$. The distribution of the sample proportion would be about normal with mean 1/2 and standard deviation $.5/10 = .05$ for samples of size 100. Hence, we would expect the sample proportion to be between .45 and .55 (within one standard deviation of the mean) 68% of the time, and between .4 and .6 95% of the time. (Recall our discussion of normal data in chapter 4.)

$$.5 + .05 = .55 \text{ and } .5 - .05 = .45$$
$$.5 + 2(.05) = .5 + .1 = .6 \text{ and } .5 - 2(.05) = .5 - .1 = .4$$

For our second example $P = 1/10 = 10\%$. Hence, the sample proportion has an approximately normal distribution with mean 1/10 and standard deviation .3/10 = .03 = 3% for samples of size 100. (Note that .3 comes from Appendix C.) We then expect that 95% of the time we will get between 4% and 16% 1s in simple random samples of size 100 from a dichotomous population with 1/10 1s.

$$10\% + 2(3\%) = 10\% + 6\% = 16\%$$
$$10\% - 2(3\%) = 10\% - 6\% = 4\%$$

Recall our discussion in chapter 4 about the number of heads that will appear in 1,000 flips of a coin. We stated then that about 500 heads would appear, or heads would appear about 1/2 of the time. Think of flipping a coin as randomly selecting an object from a large dichotomous population with 1/2 1s. Selecting a 1 is equivalent to flipping a coin and getting a head. The proportion of heads in a simple random sample selected from such a dichotomous population has an approximate normal distribution with mean 1/2 (the proportion of 1s in the population) and standard deviation of about .016. (From Appendix C the standard deviation of the population is 1/2. From table 4 of Appendix B the square root of 1,000 is 31.6, and .5 divided by 31.6 is .016.)

From the discussion of dichotomous data in chapter 4 we see that the proportion of 1s in a sample of size 1,000 is about 1/2. The proportion will be between .468 and .532 95% of the time;

$$.5 - 2(.016) = .468$$
$$.5 + 2(.016) = .532$$

that is 95% of the time between 468 and 532 1s will appear in samples of size 1,000 drawn from this population. Analogously, 95% of the time that you flip a coin 1,000 times between 468 and 532 heads will appear — that is, **about** 500 heads will appear.

Recall, also, our discussion of the gambler's fallacy in chapter 4. We asked if equilibrium (about 1/2 heads) would occur in 20 or even 100 flips of a coin. For a large number of flips of 20 coins between 6 and 14 heads in 20 flips of a coin will occur 95% of the time. Likewise, between 40 and 60 heads will likely occur in 100 flips. We see that expecting 1/2 of the flips of a coin to result in heads does not necessarily mean that a series of heads is likely to be followed by a series of tails.

## Sampling Distribution of the Mean

Just as with the proportion of 1s in a random sample drawn from a dichotomous population, we are often interested in the mean of a

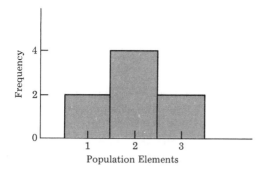

**FIGURE 7**
**A Graph of (1, 1, 2, 2, 2, 2, 3, 3)**

random sample drawn from an arbitrary population. Recall that the proportion of 1s in a sample from a dichotomous population of 0s and 1s *is* the mean of the sample. Our discussion in the previous section is then a special case of the present discussion. Here we will look at the mean of a random sample drawn from an arbitrary — not necessarily dichotomous — population.

As was the case when we looked at the sampling distribution of the proportion, we will be considering what happens when a large number of random samples is selected from a population. For each sample we calculate the mean. We want to describe this collection of means — that is, what is the distribution, the mean, and the standard deviation of this collection of means. The distribution of interest is the *sampling distribution of the mean,* or the *distribution of the sample mean.*

Two examples should help to illustrate how the mean of a sample varies from random sample to random sample. Let us suppose our population is symmetric and contains 8 elements. In particular, consider the population that has the elements.

1, 1, 2, 2, 2, 2, 3, 3

A graph of these data would look like the graph shown in figure 7.

Table 9 summarizes the results if simple random samples (without replacement) of size 2 are selected from the numbers (1, 1, 2, 2, 2, 2, 3, 3) and the means of the samples calculated. We could calculate the probability of getting each of the samples listed in table 9 or, equiva-

**TABLE 9    The Means of Samples of Size 2 from (1,1,2,2,2,2,3,3)**

| Sample | Mean |
| --- | --- |
| (1,1) | 1 |
| (1,2) | 1.5 |
| (2,2) or (1,3) | 2 |
| (2,3) | 2.5 |
| (3,3) | 3 |

lently, the chances that each of the corresponding values of the mean will occur. For example, there are a total of 6 samples of size 2 that can be selected from these 8 elements.

$$(1,1), (1,2), (1,3), (2,2), (2,3), (3,3)$$

Of these 6 samples (2,2) and (1,3) have a mean of 2. The probability that the mean is 2 is the chance that the sample is (2,2) or (1,3). This probability can be shown to be 5/14.

Figure 8 is a graph of the values of the sample mean with the corresponding probability of each such value. The sampling distribution of the sample mean for simple random samples of size 4 drawn from the population (1, 1, 2, 2, 2, 2, 3, 3) can be similarly determined.

We can see that the distribution of the sample mean is symmetric about the mean of the population which is

$$\frac{1 + 1 + 2 + 2 + 2 + 2 + 3 + 3}{8} = \frac{16}{8} = 2$$

both when the samples are of size 2 and when they are of size 4. (Recall that the mean of a symmetric collection of numbers equals the mode and the median and is the point of symmetry.) For any symmetric population, the distribution of the sample mean is also symmetric and centered at the population mean. Further, the sampling distribution of

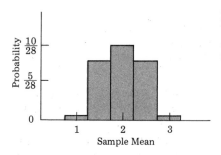

**FIGURE 8**
**The Sampling Distribution of the Mean of Samples of Size 2 Drawn from (1,1,2,2,2,2,3,3)**

TABLE 10   The Means of Samples of Size 4 from (1,1,2,2,2,2,3,3)

| Sample | Mean |
|---|---|
| (1,1,2,2) | $^6/_4$ |
| (2,2,2,1) or (2,1,1,3) | $^7/_4$ |
| (2,2,2,2) or (1,1,3,3) or (2,2,1,3) | $^8/_4 = 2$ |
| (2,2,2,3) or (2,3,3,1) | $^9/_4$ |
| (2,2,3,3) | $^{10}/_4$ |

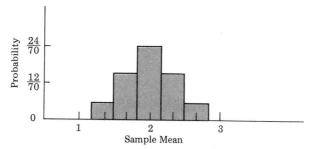

FIGURE 9     The Sampling Distribution of the Mean of Samples
             of Size 4 Drawn From (1, 1, 2, 2, 2, 2, 3, 3)

the mean for samples drawn from any large symmetric population is
closely approximated by the normal distribution. The center of this
normal sampling distribution is the population mean.

The standard deviation of the sampling distribution of the mean
is the population standard deviation divided by the square root of the
sample size.

The standard deviation of the population (1, 1, 2, 2, 2, 2, 3, 3) is
.7. The standard deviation of the distribution of the sample mean for a
sample of size 2 is .7 divided by the square root of 2(1.4).

$$\text{standard deviation of the mean} = \frac{.7}{1.4} = .5$$

Similarly, for simple random samples of size 4 drawn from (1, 1, 2, 2, 2,
2, 3, 3) the standard deviation of the sampling distribution of the mean
would be .35.

$$\text{standard deviation of the mean} = \frac{.7}{2} = .35$$

## When the Population Is Not Symmetric

As we discussed in the preceding section, when the population is
symmetric and large, the normal distribution is a good approximation
of the distribution of the sample mean. If a population is large but *not*
symmetric, one wonders what the sampling distribution of the mean
might look like. Consider the skewed (to the right) collection of numbers

1, 1, 1, 1, 1, 2, 2, 3

If simple random samples of size 2 are selected from this skewed population and the sample means calculated, we would find the results

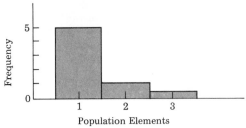

**FIGURE 10     A Graph of (1, 1, 1, 1, 1, 2, 2, 3)**

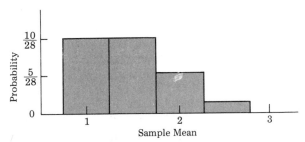

**FIGURE 11     The Sampling Distribution of the Mean of Samples of Size 2 Drawn From (1, 1, 1, 1, 1, 2, 2, 3)**

summarized in figure 11. Figure 12 shows the sampling distribution of the mean for samples of size 4.

The mean of the sampling distribution of the mean is 1.5 for both samples of size 2 and for samples of size 4. The mean of the population is also 1.5. (Observe that with the skewed distribution in figure 11 [samples of size 2] the mean of 1.5 is not the "center" of the distribution. For the symmetric distribution of figure 12 [samples of size 4] the mean of 1.5 is the center of the graph or the point of symmetry.) The standard deviation of the sampling distribution of the mean for samples of size 2 is .7; for samples of size 4 it is .35.

The mean and standard deviation of the sampling distribution of the sample mean are found for the skewed population just as they were for symmetric population. The only difference in these sampling distributions are their shapes. For samples of size 2 the sampling distribution of the mean is skewed to the right just as the original population (1, 1, 1, 1, 1, 2, 2, 3) was skewed to the right. However, for samples of size 4 the sampling distribution of the mean has become symmetric, even though the population from which we are selecting samples is skewed. (We do not have to have a sample size equal to one half of the population size for this to occur.) Generally, for large skewed populations the sampling distribution of the mean will become symmetric and nearly normal for large sample sizes. The center of this normal

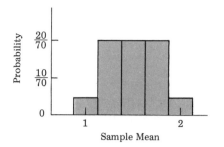

FIGURE 12
The Sampling Distribution of
the Mean of Samples of Size 4
Drawn From (1,1,1,1,2,2,3)

curve is the population mean. The standard deviation is given by the population standard deviation divided by the square root of the sample size.

## Central Limit Theorem

We have seen from our discussion in this and the preceding sections that the sampling distribution of the mean and proportion are nearly normal. The larger the sample size the closer to normal are these sampling distributions. The statement that the sampling distributions of the mean and proportion approach a normal distribution is called the Central Limit Theorem.

> *Central Limit Theorem*  If random samples are drawn from a large population, the distribution of the sample mean will be about normal. The center of this normal curve is the population mean, and the standard deviation is the population standard deviation divided by the square root of the sample size. The larger the sample size the closer the distribution of the sample mean is to being normal.

Although we have talked about simple random samples, the sampling distribution of the mean is closely approximated by the normal distribution for the other *random* sampling designs discussed in this chapter. We cannot go too far wrong by assuming normality of the sampling distribution of the mean (proportion) when trying to measure the uncertainty of the resulting statistical inference. We can usually assume that the sample sizes are large enough for such an approximation. (For continuous collections of data, 30 is usually considered a large enough sample size. A sample size of less than 30 would be sufficient when the population is symmetric, although more than 30 observations will be needed for skewed populations.)

# Summary

We have explained why preelection polls for the 1936 and 1948 presidential elections might have been wrong. In 1936, working from an inadequate frame, or population list, the *Literary Digest* used a voluntary response straw vote when they incorrectly predicted an electoral college landslide for Landon over Roosevelt. In 1948 pollsters failed to detect late shifts in the "undecided" respondents, causing them to incorrectly predict Dewey over Truman.

After discussing what can and has gone wrong with surveys, we considered the different sampling scientific procedures that might be used to select a sample:

1.  Simple random sample.

2.  Stratification techniques are widely used to help ensure a representative sample. Some form of stratification is used in almost every survey run today. (Pre)Stratification is the dividing of a population into parts called strata. Samples may then be drawn of each stratum. In poststratification the sample is divided into strata. Weighting of sample data by known sizes of strata may then be carried out. Caution is urged in this regard.

3.  Systematic samples, as used by the Census Bureau when sample surveys are run simultaneously with a census, require the selection of every so many elements of a population from a random starting point.

4.  Cluster sampling, often one stage of a multistage selection procedure, involves the selection of groups or clusters of population elements. For example, city blocks, which are clusters of households, might be selected.

Having discussed how to select a sample, we looked in some detail at a special inferential statistic — the sample mean. (The proportion of 1s in the sample is an important special case of a sample mean. The sample proportion is the mean of the 0s and 1s in a sample selected from a dichotomous population.) Realizing that the value of the sample mean necessarily changes from random sample to random sample, we asked if this sample to sample change or distribution is predictable. The Central Limit Theorem tells us that for most large populations this sampling distribution of the mean is approximately normal and that this approximation improves with an increase in sample size. The center of the normal sampling distribution of the mean is the population mean. The standard deviation of this normal distribution is the standard deviation of the population divided by the square root of the sample size. In the next chapter we will see how this information can be used in practical situations when the parameters in question, the mean and standard deviation of the population, are not known.

Recall at this point in our discussions some comments made in the last chapter. We commented that the uncertainty of statistical inference would be measured by normal probabilities. This must have seemed quite strange at the time, since the population we might want to investigate may not be normal. Indeed, a dichotomous population is far from normal. We have built upon this idea in the latter sections of this chapter. We *can* measure the uncertainty of statistical inference using normal probabilities. The normality comes from the statistics not from the population. The change of certain inferential statistics from random sample to random sample is predictable. For example, the values of the sample mean vary from random sample to random sample in a normal pattern! It is this normal distribution, the sampling distribution of the mean, that gives us a (normal) measurement of the uncertainty of statistical inference.

*Note:* Having been interviewed for a Gallup survey and having kept a diary for a radio survey, I felt that these experiences might be of interest to you. As an aside, you might want to turn to Appendix A for a discussion of my experiences. You will likely get a better idea of opinion polling and radio- (and television-) viewing surveys.

# No Comment

■ *"How the Survey Worked"*

 *"Letters were mailed to all 2,696 delegates and alternates to the 1972 convention in Miami Beach, Fla., that nominated Richard Nixon and Spiro Agnew. About 300 of those letters (in which we explained our reasons for undertaking this survey) were not delivered, because the delegates had died, moved, or simply because the mail didn't get through. Of the people we contacted, 789 returned the stamped, addressed post cards we supplied. Of these returns, 37 could not be tabulated. Finally, then, 752, or 31.4 per cent of the people we contacted took part in the survey."*

       *(Survey conducted by the* National Observer)

■ *"The telephone survey was not based on a scientifically selected sampling of telephone customers."*

       *(Survey conducted by* Lexington (Ky.) Herald)

■   *"Problem — Now, are the figures in this Poll correct? In answer to this question we will simply refer to a telegram we sent to a young man in Massachusetts the other day in answer to his challenge to us to wager $100,000 on the accuracy of our Poll. We wired him as follows:*

*"For nearly a quarter century, we have been taking Polls of the voters in the forty-eight States, and especially in Presidential years, and we have always merely mailed the ballots, counted and recorded those returned and let the people of the Nation draw their conclusions as to our accuracy. So far, we have been right in every Poll. Will we be right in the current Poll? That, as Mrs. Roosevelt said concerning the President's reelection, is in the 'lap of the gods.'*

*"We never make any claims before election but we respectfully refer you to the opinion of one of the most quoted citizens to-day, the Hon. James A. Farley, Chairman of the Democratic National Committee. This is what Mr. Farley said October 14, 1932:*

*" 'Any sane person can not escape the implication of such a gigantic sampling of popular opinion as is embraced in THE LITERARY DIGEST straw vote. I consider this conclusive evidence as to the desire of the people of this country for a change in the National Government. THE LITERARY DIGEST poll is an achievement of no little magnitude. It is a Poll fairly and correctly conducted' "*

*(From the* Literary Digest *concerning their 1936 poll)*


■   *"The election outcome shocked pollsters who predicted the 54-year-old Liberal leader would not win 133 seats in the House of Commons."*

*(Newspaper report of Pierre Elliot Trudeau's 1974 election victory in Canada)*


# Exercises

1.   The method of randomly generating telephone numbers for a sample survey (see, for example, A Gallup Interview" in Appendix A) has some advantages and disadvantages over selecting numbers from a telephone directory. What are they?

2.   Select a simple random sample of, let's say, size 5 from your class so that you can get a sense of random selection techniques.

3.   The student government at a university selected a random sample of students who were to serve on the judicial board by putting the name of each student into one of 26 containers accord-

ing to the first letter of a student's last name and randomly selecting one name from each box. Is this a simple random sample of size 26? What is the stratification technique used? Is it reasonable? Convenient?

4. An ex-senator described his survey of citizens' opinions:

> Each June, I send the questionnaire to all the individuals and households on my newsletter mailing list. This year that amounted to 208,300; of that number, just about 22,000 returned the form. As they came into the office the questionnaires were counted, the number logged and the forms kept separate by day. When the number being returned fell to just a few daily, 10% of each day's questionnaires were pulled at random and the responses tallied. From that sample the statistical analysis shown in the enclosed newsletter was made.

What type of random sampling plan is being used here? To what population(s) can inference be properly made?

5. Suppose that a poll indicates 45% of a population favor a candidate. If 1,000 people were interviewed, what is the standard deviation of the above estimate — that is, what is the standard deviation of the sampling distribution of this statistic? The sample proportion is likely to take on what values? Suppose 1,000 people were selected from a county of size 200,000, or a state of size 3 million, or from the whole country — would your answer change?

6. A census tract should not contain both expensive and slum housing as averages would not properly reflect either group. What might be the distribution of data from a census tract containing both slum and expensive housing? How could central tendency be represented?

7. We mentioned that the sample size for most surveys is large enough for good normal approximation of the distribution of the sample proportion of 1s. Although this may be true for the total sample, what if someone is interested in making statistical inference using, perhaps, all the females between the ages of 40 and 50 years who were in a sample? Is this restricted sample large enough?

8. Consider the distribution of the sample proportion when taking samples of size 1,100 from a dichotomous population with 36% 1s. The standard deviation of the population is (see Appendix C) about .5, and the square root of 1,100 is 33.2 (see Appendix B,

table 4). The distribution of the sample proportion is then approximately normal, with mean .36 and standard deviation of .015 (.015 = .5/33.2). What are the chances that we will select a simple random sample with between 34.5% and 37.5% 1s? What would you think if someone said that they took a random sample from this dichotomous population and got more than 40.5% 1s?

9.  Scores on the first exam in a statistics class had a mean of 50 and standard deviation of 10. Four students were randomly assigned to work on a joint project. Other students in the class complained that these 4 students had an unfair advantage since they averaged 75 on the first exam. The other students felt that these 4 students couldn't have been selected randomly. If the teacher had in fact assigned students to project committees in a random manner, what are the chances of getting a committee of 4 that averaged at least 75 on the first exam? Do you buy the teacher's claim that committee assignments were made randomly?

10. Jimmy Carter received 51% of the vote in the 1976 presidential election. For a sample size of 3,600, what would you expect the distribution of the sample proportion to be for the preelection polls run just prior to the election? What would you think of a pollster's estimate that Carter would get 48% of the vote.

   "Other" candidates got about 1% of the vote during this election. What kind of a job did a pollster do who predicted a 3% vote for the other candidates (Eugene J. McCarthy, Lester G. Maddox, etc.)? (Assume a sample size of 3,600.)

# For Discussion

1.  Discuss my interviews by the Gallup organization, which are presented in Appendix A. How representative of public opinion is such a poll?

2.  Suppose that a credible scientific sample is selected from a population. Does this guarantee accurate survey results?

3.  Discuss the difficulty of getting a list-type frame from which to select a random sample for opinion polling, preelection surveys, and television-viewing and radio-listening surveys. What might a frame be in these instances?

4.    Some random samples selected from your class (see Exercise 2) may not represent the class according to all criteria. Why not just pick a quota of 5 students then?

5.    Discuss the cost involved in getting a list of elements from different populations. Is cluster sampling likely to save a lot of money?

6.    Why might an interviewer be asked to take a systematic sample of households from a block, rather than a simple random sample of households?

7.    Discuss the selection of a representative sample of students from your school. Include a discussion of the sample of university students discussed in the section called "The Sample Survey" in chapter 2. A systematic sample of numbers from the school's student telephone directory was selected. About 100 student interviews were completed. Is such a sample survey of students likely to accurately reflect the opinion of all students?

8.    The sampling of a wildlife population is frought with many problems. Discuss the problem of selecting a sample from, let's say, a whale population. (See, for example, D. G. Chapman, "The Plight of the Whales," *Statistics: A Guide to the Unknown,* ed. J. M. Tanur [San Francisco: Holden-Day, 1972], pp. 84–91.)

9.    Preelection polling for the 1968 election is an interesting example to consider. Following the chaotic Democratic convention in Chicago, the Democratic nominee, Hubert Humphrey, trailed the Republican candidate, Richard Nixon, by 16%. By late September the margin had changed little. By early October the gap narrowed to 12 percentage points. The third week of October revealed the difference to be 8%. The last poll, reported the night before the election, reported the election too close to call. (The final margin was less than 1%.) Discuss how important trends are in preelection polls. Contrast these preelection polls with the 1936 and 1948 experiences.

10.    Take a look at a map of census tracts for a SMSA near you. Discuss the use of such a map in designing a survey of this metropolitan area. (We'll look at survey design in chapter 8.) The planning commission in the city chosen should have such a map. Otherwise, the Census Bureau has them.

11.   Discuss the cost of auditing all the accounts in a bank as compared to auditing a sample of accounts. Would you stratify? What about allocation of the sample to the strata?

12.   Discuss the audience share defined in "An Arbitron Diary" in Appendix A. (Recall our discussion of the influences of definitions on the interpretation of statistics. A local public television station got an audience share of 100% for the program "Man: The Incredible Machine." This number is a form of doublespeak unless one realizes that an electrical storm put all commercial stations out of commission that evening). With regard to the audience share, what demand of the sample size of television-viewing surveys might result from a need to know this number?

13.   Recall in the section on the sample survey in chapter 2 we stated that the sample size and error (standard deviation of the distribution of the statistic) were two pieces of information needed to judge the results of a survey properly. Recount the other bits of information (background influences) required by tying them to these last two important bits of information.

14.   Compare quota sampling and stratified random sampling. Concentrate on the role of the interviewer. Does the interviewer select people to be interviewed in each case? Can this adversely affect survey results?

# Further Readings

The misuse of polls in the political arena is discussed in both Roll and Cantril's *Polls; Their Use and Misuse in Politics* and Wheeler's *Lies, Damn Lies and Statistics*. Both are cited in the chapter Notes. Recall that we look at preelection polls since the results of an election tell us just how accurate each poll was. This is one area in which we can document examples of poor polls.

Gallup's *The Sophisticated Poll-Watchers Guide* gives you the pollster's side of the issue of opinion polling. This is also listed in the Notes section.

Nontechnical discussions of the sample survey can be found in Des Raj's *The Design of Sample Surveys* (New York: McGraw-Hill, 1972) and Morris Slonin's *Sampling* (New York: Fireside, 1960).

# Notes

1.   C. W. Roll, Jr., and A. H. Cantril, *Polls: Their Use and Misuse in Politics* (New York: Basic, 1972) p. 10.
2.   G. Gallup, *The Sophisticated Poll-Watchers Guide* (Princeton: Princeton Opinion, 1972).

3.  M. Wheeler, *Lies, Damn Lies and Statistics* (New York: Liveright, 1976), p. 70.
4.  F. Mosteller, *The Pre-election Polls of 1948, Report to the Committee on Analysis of Pre-election Polls and Forecasts* (Washington, D.C.: Social Science Research Council, Bulletin 60, 1949): 105.
5.  *Ibid.,* p. 52
6.  *Ibid.*
7.  *Ibid.,* p. 301.
8.  *Business Week* 13 (November 1948), p. 26.
9.  Information sent by Zenith Corporation.
10. J. Neter, "How Accountants Save Money by Sampling," *Statistics: A Guide to the Unknown,* ed., J. M. Tanur (San Francisco: Holden-Day, 1972), pp. 203–211.
11. Wheeler, *Lies, Damn Lies and Statistics,* p. 120.
12. M. S. Rath, "On Approximating the Point Binomial," *Journal of the American Statistical Association* 51 (1956): 293–303.

# 6 Statistical Inference

IN THE MID-1970s Canadian researchers revealed that saccharin produced bladder tumors in rats. Saccharin was fed to test rats at a concentration of 5% of their total diet. In order to properly ascertain the effects of saccharin on these rats a second group of rats, called control rats, was fed and housed as were the test rats with the exception that the control rats received no saccharin in their diet. Researchers noted a significantly higher incidence of bladder tumors in the second generation of test rats over the second generation of control rats. To interpret the results of this research, we need to know something about the statistical inference that led to the conclusions.

We will look at the use of the sample proportion (an inferential statistic) as an approximation of the proportion of 1s in a dichotomous population (a parameter). A consideration of the accuracy of such an estimation procedure leads to a discussion of confidence intervals, which are interval approximations of a parameter. The word "interval" refers to a range of values as opposed to an "estimate," which is a single number. We will also consider the estimation of the mean of a continuous population.

Situations in which we seek to ascertain the accuracy of a statement concerning a parameter of a population are discussed. Examples of tests of hypotheses concerning the proportion of a dichotomous population and the mean and median of a continuous population are investigated.

We will also discuss the role of descriptive statistics when inference may be of particular interest. With the difficulty that exists in trying to observe an entire population, we often encounter inferential statistics; desirable, descriptive statistics are too often not reported.

In addition we will see how probability measures the uncertainty of statistical inference. The connection between the normal probabilities discussed in chapter 4 and statistical inference will be seen. Recall from our discussion in the last chapter that the change in the sample proportion from random sample to random sample is predictable. Indeed, the sampling distribution of the proportion is about normal. Basing statistical inference on this near normal statistic allows us to measure the uncertainty of inference using normal probabilities.

Let us see the relationship between probability and statistical inference.

# Estimating a Population Proportion

Let us take a detailed look at the estimation of the proportion of 1s in a dichotomous population. Recall that the elements in a dichotomous population are classified into one of two categories.

As discussed in chapter 5, we represent all the elements in a dichotomous population by a 1 or a 0. If we are looking for the proportion of "unemployed" persons in the labor force, all such individuals are labeled 1, and all other persons in the labor force are labeled 0. Our interest in a dichotomous population will be with the proportion of 1s in the population in which 1 represents the particular category of interest.

In our discussions of proportions the words "proportion" and "percentage" are used interchangeably. A percentage differs from a proportion by a factor of 100. For example, the proportions of .8 and .75 are equivalent to 80% and 75%, respectively.

Recall that the mean of the 0s and 1s in a dichotomous population is equal to the proportion of 1s in the population. This fact will give us a natural tie-in between our work here and our discussion of inference to population means later in this chapter.

## The Estimate

In situations in which inferential statistics are to be used, the proportion of 1s in a population is **not known**. If, for example, we are interested in the proportion of "defective" items manufactured by a company, 1 would represent a defective item and the proportion of 1s is equivalent to the proportion of defective items. This proportion is not known. In order to get some idea of what this unknown proportion (a parameter) might be, a sample will be selected from the population. Methods of selecting a sample were discussed in chapter 5. We will only assume that the sample is some type of random sample.

With interest in the proportion of 1s in our population, we would intuitively want to investigate the proportion of 1s in the sample. Recall that this latter quantity is called a statistic. This statistic (the sample proportion) is inferential when we use it to make a statement about the unknown population proportion — we are **inferring** to this unknown population proportion. One type of inference is to use sample information to approximate an unknown population proportion.

An **estimate** of a population parameter is an approximation of this unknown value. The sample statistic, which corresponds to this population parameter, is the approximation used.

In the dichotomous population case, we would estimate the proportion of 1s in the population by the proportion of 1s in a sample drawn from that population — that is, if 8.5% of our sample of adults (16 years or older) in the labor force are "unemployed" (8.5% is a statistic), we estimate the proportion of *all* adults in the labor force who are "unemployed" as 8.5%. If 35% of the people in our sample who are watching television are watching the Academy Awards show, we would use this number to estimate the proportion of *all* people who are watching the Academy Awards show.

**Why a Random Sample**    Why must we go to all the trouble of taking a random sample to get an estimate of a population proportion? A talk with an officer of an unemployment office could lead to an estimate of the proportion of unemployed. A political analyst could estimate the proportion of registered voters who are likely to vote for a candidate. Or a television executive may feel that he has a pretty good idea of what the audience share of a particular show will be. Anyone can estimate a population proportion. We must therefore try to answer the question, "How good is a particular estimate?"

Unless we know how close an estimate tends to be to the unknown parameter, we have no way of judging how good an estimate is or which of several estimates is best. It is often very difficult to judge how close an estimate is to the parameter. A guess at the unemployment rate or at an audience share or at a future vote is just that — a guess! No one knows how close their estimate might be to the true population proportion.

But all is not lost. The reason that a statistic from a random sample is a desirable means of estimating a population parameter is that a measure of closeness of the estimate to the unknown parameter is available — that is, the relationship between the estimate and the parameter can be measured using the tool of probability inherent in the random selection process.

**The Relationship between Population Parameter and Its Estimate**
What, then, is the relationship between the population proportion and the proportion from a random sample — our estimate of the population proportion?

From our discussions in chapter 5 we recall that the sample proportion varies from random sample to random sample in a predictable manner. The value of the sample proportion will necessarily change with each distinct sample that is drawn from a dichotomous population. In particular, values of the sample proportions from different samples follow a near normal pattern. The center of this normal curve is the population proportion. It does not matter if this population proportion is known or unknown.

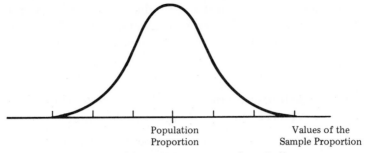

| | |
|:-:|:-:|
| Population<br>Proportion | Values of the<br>Sample Proportion |

**FIGURE 1**   Values of Sample Proportions from Different
Random Samples

Figure 1 graphically illustrates the distribution of values of sample proportions from different random samples. The height of the graph represents probability. By "represent" probability we mean that the probability associated with intervals on the data scale is the area under the curve bounded by the interval. Recall our discussions in chapter 4 regarding the normal distribution.

The sample proportion is just as likely to be too large an estimate, above the population proportion, as too small an estimate, below the population proportion — that is, the distribution of the sample proportion is symmetric about the population proportion.

**Standard Error**   As we have discussed, the sampling distribution of the proportion is approximately normal. The mean of this normal distribution is the population proportion. The standard deviation of the sampling distribution of the proportion is also called the ***standard error of the proportion***. Recall then that

$$\text{standard error of the proportion} = \frac{\text{standard deviation of the dichotomous population}}{\text{square root of the sample size}}$$

Using our knowledge of a collection of normal values, we observe that the sample proportion will be within two standard errors of the population proportion for 95% of samples we might select or with 95% probability. This statement can be turned around to say:

> The population proportion will be within two standard errors of the sample proportion with 95% probability.

Recall from our discussion in chapter 5 concerning the distribution of the sample proportion that the standard deviation of this sampling distribution, the standard error of the proportion, depends on the standard deviation of the dichotomous population, which in turn depends on the population proportion. The population proportion is, however, not known. We must therefore estimate the standard error of the proportion.

$$\frac{estimated \text{ standard error}}{\text{of the proportion}} = \frac{\text{standard deviation of the } sample}{\text{square root of the sample size}}$$

Observe that the estimated standard error involves the sample standard deviation; the actual standard error involves the population standard deviation.

For example, suppose a sample of size 100 results in 64 1s. We would *estimate* the population proportion to be .64. If the population proportion were .64, the standard deviation of the population would be .5 (see Appendix C). We estimate the standard error of the proportion to be

$$\frac{\text{sample standard deviation}}{\text{square root of sample size}} = \frac{.5}{10} = .05$$

Observe that if the population proportion is .64, the standard error of the proportion is .05. Since we only estimate the population proportion to be .64, .05 is an *estimate* of the standard error of the proportion. For this example, the sampling distribution of the proportion would be approximately normal as shown in figure 1, with an estimated standard error of .05.

Suppose, as a second example, that 500 elements are selected with 248 1s. This sample proportion is

248/500 = .5

Thus, .5 is our estimate of the proportion of 1s in the population. The estimated standard error of the proportion is

.5/22.3 = .022

Observe that .5 comes from Appendix C and 22.3 is the square root of 500 (see Appendix B, table 4).

**A Definition of Probability Modeling**    We should think of the structure we have described as probability modeling. We have modeled the sampling process using a normal distribution to describe the change in the sample proportion from sample to sample. This model affords us the opportunity to measure the uncertainty of statistical inference, as we will see. But how well does this model fit real-life sampling situations? Even very poorly run surveys, surveys with poor response rates, or surveys with voluntary rather than random selections will assume this model. In these cases or, in general, when a consideration of background influences suggests a poorly run survey, this model does not fit the real-life situation and will give a distorted measure of the uncertainty inherent in statistical inference. We may not be able to measure uncertainty in such cases. For properly designed and properly executed surveys with a random sampling procedure the normal model fits quite well, as long as the sample size is large (see chapter 8 regarding properly designed surveys). One exception is cluster sampling for which the standard error mentioned in our probability model may be too small. Another exception is continuous data. In this case the standard error can change dramatically with changes in the sampling procedure. Generally, when estimating the proportion, the normal model we have described is adequate for properly designed and properly executed random sample surveys.

### Interpretations

As we stated earlier the population proportion is within two estimated standard errors of the sample proportion with 95% probability. For example, if a preelection poll finds that 64% of 100 registered voters interviewed intend to vote for a particular candidate, then we can be confident that between 54% and 74% of all registered voters favor this candidate. Note that the standard error is 5% and that

$$64\% - (2 \times 5\%) = 54\% \text{ and } 64\% + (2 \times 5\%) = 74\%$$

That is, percentages within two standard errors of 64% are between 54% and 74%.

In the preceding example, is the true proportion of all registered voters who favor the candidate between 54% and 74%? Of course we do not have the answer to this question since the proportion of all registered voters favoring the candidate is not known; so, who can say whether this percentage is between 54% and 74%? (Recall that only by talking to all registered voters can we find the unknown proportion we seek.) Then, how is the interval between 54% and 74%, called a *95% confidence interval for the population proportion*, to be interpreted?

There is not a 95% chance that the unknown population proportion is between 54% and 74%. The unknown proportion *is* between .54 and .74 or it *is not* — there is no probability involved. Our statement, "With 95% probability the population proportion is within two standard errors of the sample proportion," reflects our confidence in the *process* of taking random samples and obtaining intervals. Ninety-five percent of the random samples that we can draw from our dichotomous population will result in an interval containing the unknown population proportion. Five percent of the random samples we might draw will result in intervals not containing the unknown population proportion. We don't know if the sample interval we actually obtained (54% to 74%) is one of the 95% accurate intervals or one of the 5% inaccurate intervals. Since we are 95% sure that our process will give us an accurate interval, we talk of our interval as a 95% "confidence interval."

> A *95% confidence interval* for a population proportion consists of all values within two estimated standard errors of the sample proportion.

**Interval Size**    We must be careful when considering the size of confidence intervals for unknown population proportions. The size of a confidence interval is the difference between the interval limits. The size of a confidence interval is usually expressed in terms of percentages. The size of an interval for the true population proportion could give us a false feeling that the interval is narrow — that is, we might feel we are close to the unknown population proportion or percentage. But percentages are based on the size of the population. For example, suppose it is found that 20% of the households in a state view public television. For a sample of size 1,000, a 95% confidence interval would be 17.5% to 22.5% of the households in the state under consideration.

$$20\% + 2\,(1.25\%) = 20\% + 2.5\% = 22.5\%$$
$$20\% - 2\,(1.25\%) = 20\% - 2.5\% = 17.5\%$$

This interval appears small, since it is only 5 percentage points in length

$$22.5\% - 17.5\% = 5\%$$

You might get an inappropriate feeling of accuracy in this case. For the state in question, the interval in terms of households would be from 175,000 viewing households to 225,000 viewing households — a difference of 50,000 households. Expressing the interval in the latter manner makes it not seem as accurate an assessment of the number in the viewing audience.

For another example of how impressions may vary relative to how the size of confidence intervals is expressed consider the unemployment rate of 8.5% for February 1975. (Recall Discussion 10 of chapter 2.) The government interviews about 47,000 households, which would be about 105,750 people who are at least 16 years old. (There are about 2.25 persons 16 years or older per household in the

United States.) A 95% confidence interval would be approximately 8.328% to 8.672% — an interval of length .344%. If, on the other hand, we were talking about the number of unemployed persons (rather than the rate of unemployment), this "small" interval "grows." With a labor force of about 93,564,000 in early 1975, we get an interval of 7,792,010 to 8,113,870 unemployed people — a range of error of 321,860 people (the approximate size of Des Moines, Iowa).

(Here our model may not be appropriate. The government interviews people on a continuing basis. This means that only a portion of the people are being interviewed for the first time; others may have been interviewed many times before. This is called sampling on successive occasions. We are, however, probably not too far off assuming the model we have chosen.)

We have discussed the error involved in estimating a population proportion by using a sample proportion. Be aware that although our estimate of a percentage might seem accurate, the size of the population might be such that a small variation as measured in percentages can be a large variation when measured by the number of population elements.

The size of the interval going from percentages to the number of population elements changes by the size of the population. For example, the .344% interval for unemployment mentioned in our last example represents $.00344 \times 93,564,000 = 321,860$ people.

**Other Sources of Error**    Again, we must be cautious. The standard error of an estimate may not be the only source of error. Recall from our discussions in chapter 2 that background influences may affect the accuracy of our inference. The errors reported in this section are assuming that a good random sample is selected and is properly and completely analyzed. If there is a high nonresponse rate for a survey or some other undesirable influence on the results, the difference between the sample proportion and the population proportion can be much greater than indicated by the error estimates given in this section, that is, the model we have proposed is not appropriate. For example, a sample proportion might vary by 20% or 30% from the population proportion when the standard error is only 3%. The point is that the standard error may be but a small part of the *overall* error if statistics are not properly obtained. We will not know the overall error rate for improperly collected data, so it makes little sense to talk about the standard error in such cases. Numbers obtained from improperly collected data are meaningless as there is not likely to be an adequate model to describe the change in the sample proportion from sample to sample. Meaningless numbers could result in doublespeak, as discussed in chapter 1, if we don't realize the numbers are meaningless.

# Testing a Statistical Hypothesis
# Concerning the Population Proportion

In the last section we were concerned with estimating an unknown population proportion. We estimated the population proportion (a parameter) by the sample proportion (an inferential statistic). An interval estimate takes into account how the sample proportion varies from random sample to random sample. For a random sampling procedure this variation of the sample proportion is predictable and gives us a more meaningful idea of the value of the unknown population proportion.

In this section we will consider an entirely different type of statistical inference — the test of a statistical hypothesis.

> A *statistical hypothesis* is a statement concerning the distribution of the numbers in a population. A statistical hypothesis is, for our purposes, a statement about a parameter of a population.

In the case of a dichotomous population, the distribution is completely specified by the proportion of 1s in the population so that a statistical hypothesis in this case is a statement about the population proportion. A testing situation arises when we are interested in the validity of a statistical hypothesis — that is, when we are interested in the validity of a statement about a population proportion. In contrast, for the estimation situation of the preceding section we were not interested in the validity of a statement about the population proportion. In the preceding section we were interested in approximating an unknown population proportion.

There are many examples of testing situations. On the night of a presidential election we listen to a projection of who will win. All three national networks, ABC, CBS, and NBC, spend much money in an effort to be the first network to predict the winner accurately. The statistical hypothesis of interest here is whether a candidate will get a majority of the votes cast; that is, is the proportion voting for a particular candidate greater than 1/2? A drug company may ask whether the level of a chemical in a drug is within an acceptable range. A low level of the chemical may mean that the drug would be ineffective, and a high level may mean that the drug would be harmful. A politician may wonder whether more than 50% of his constituents favor a particular piece of legislation.

Other examples of statistical hypotheses involve two (or more) dichotomous populations. Our interest here is whether the two populations are identical, that is, whether the proportions of 1s in the two dichotomous populations are the same. For example, we might ask whether the populations sampled by two different survey organizations are the same. (Recall that two surveys might be targeted to the same population but differences in the timing of the surveys, different methods of contact — personal interview or telephone — or different response rates for the surveys may make the populations actually sampled quite different.)

Two different drugs might be tested to see whether the proportions of people who are "satisfied" with these drugs are equal. (Satisfaction could be the lack of stomach upset, relief of pain, or a cure.) If a difference is found, we are likely to hear about it in an ensuing advertising campaign: "Our product was found to be *significantly* better than (the infamous) Brand X." Such claims of significance are quite common. Examples include: "[by using an analgesic rub] in many cases the gripping strength was significantly improved," and, "[for a drug containing caffein] results were significant — drivers maintain alertness better."

The importance of the word "significance" in advertising was illustrated by a 1973 Federal Trade Commission (FTC) suit against three manufacturers for advertising claims that their aspirin products were significantly superior to others. The FTC contended that products that have aspirin as the major ingredient cannot differ significantly. Hence, the FTC proposed that the companies involved should not advertise their products for two years unless 25% of their advertising expenses were used in corrective ads. Such corrective ads might be, "All aspirin is about the same, but ours. . ."

Let us take a detailed look at two examples of tests of statistical hypotheses. We will consider the projection of a winner on election night as well as a test that the results of two surveys are the same.

### Test of a Hypothesis: One Population

Projecting a winner of an election after the polls have closed but before all the votes have been counted has become a widespread practice. A colleague of mine, we'll call him Lynn, got into the act. (This is an area where a statistician treads carefully since the actual vote will be known by the next morning.) Working for a local radio station, Lynn selected a stratified random sample of precincts from our metropolitan area. Representatives of the radio station went to these key precincts. When the polls closed and the precinct totals were posted, workers at the key precincts called in the results, which were then analyzed. On the basis of the results of these key sample precincts the winner of the election was predicted and announced over the radio. Lynn's first attempt to project winners was a hard act to follow because not only were the mayoral projections correct but council races were also predicted with extreme accuracy.

The next election was a different story. The wrong candidate was projected the winner. (Protests, recounts, and much hoopla reversed the results some weeks later, but by then the [statistical] damage had been done.) The radio station had taken matters into its own hands and had selected its own sample of precincts. These precincts were not, however, randomly chosen. Precincts were chosen that had been close to the actual vote in the previous election. These precincts failed to vote the "right way" in this election and the radio station's attempt at projecting a winner was a disaster.

Precincts can be selected because they seem to always vote for the winner. We say that these precincts are good predictors of the actual vote. Observe that these precincts do not have to be a random sample of precincts, just good predictors of the actual vote. The analysis of this type of data gets involved with a prediction problem in which the estimate of the actual vote and the error of this estimate are different than that given in the first section of this chapter. (A different model is assumed.) See chapter 9 for a look at the prediction (regression) problem. Link has a discussion of predicting the vote using predictor precincts instead of a random sample of precincts.[1]

It is interesting to note that what happened in our city is what can be expected at the national level. National news networks choose precincts across the country. Workers call in precinct results as soon as the polls close. Members of the League of Women Voters have, for example, been used as precinct workers by ABC news.

Much preparation goes into the few minutes of work done by the workers at the key precincts on election night. A telephone may have to be installed for the all-important election-night call. Precinct workers talk to voting officials so that results can be obtained quickly. (The use of voting machines is very helpful here.) A trial run is even made before election day to assure that all will go well on election night.

Often all does go well. As little as 10 minutes after polls close in a state, the projection of a winner might be heard from the news media. For example, Kentucky went for Nixon in 1972. The projection came at 7:10 PM EST, 10 minutes after polls closed in that state! In 1976 Indiana and Kentucky were projected to go for Ford and Carter, respectively, at 6:30 PM EST, before the polls closed in the western parts of these states! (The western parts of these states are on central standard time, although the eastern parts are on eastern standard time.)

But sometimes, as with our local radio station, things do not go very well. In the 1976 Wisconsin Democratic presidential primary, ABC and NBC predicted Congressman Morris Udall the winner. Udall even gave a long-awaited victory speech. But by 3:00 the next morning it became clear that Carter had in fact won, leaving CBS executives and Carter smiling, while ABC, NBC, and Udall scrambled for explanations.

**The Test**  Let us look in detail at the problem of deciding who will win an election before all the votes are counted. We are interested in which candidate will get a majority of the votes. Assume there are only two candidates (the dichotomous case); so, the question is who will get more than 50% of the vote. Our statistical hypothesis will involve the proportion of votes for, let's say, Candidate A. We ask if this proportion is less than 1/2 (Candidate A loses) or greater than 1/2 (Candidate A wins). We test these possibilities indirectly by testing the hypothesis that the proportion of votes cast for Candidate A equals 1/2. (Our intuition might be aided by thinking of this hypothesis as the proportion of votes cast for Candidate A is "about" 1/2 — a close election.) The hypothesis that we are to test involves a statement that a population parameter *equals* some value. We will see later why this is necessary.

---

The statistical hypothesis that is to be tested is called the *null hypothesis*.

---

We are testing the null hypothesis that the proportion of votes to be cast for Candidate A is 1/2. Alternatively, this proportion may be greater than or less than 1/2, in which case we, the network executives, should project the appropriate winner.

We are therefore confronted with a two-action decision problem:

*Action 1*: Fail to reject the null hypothesis that Candidate A gets 1/2 the votes, thereby airing the familiar, "This election is too close to call."

*Action 2*: Reject the null hypothesis that Candidate A will get 1/2 the vote, and project a winner. We project Candidate A if we feel the proportion in question is greater than 1/2; we project A's opponent if the proportion is thought to be less than 1/2.

A few words are in order concerning the use of the phrases "reject the null hypothesis" and "fail to reject the null hypothesis." We do not "accept" the null hypothesis as suggested in some statistic texts. As an illustration, consider the hypothesis "This man is poor." We would have to know the man in question quite well before we could "accept" the fact that he is poor. On the other hand one strong piece of information could cause us to "reject" this hypothesis. For example, if we found out that the man in question had $10,000 in a savings account, we would reject the idea he is poor. Other information we might find out about this man might include that he walks to work and carries his lunch. These two facts will not allow us to accept the fact that this man is poor, nor will we be able to reject this idea. We would say that the fact that this man walks to work and carries his lunch leads us to "fail to reject" the hypothesis that he is poor.

In our election-night projection example we are testing the statistical hypothesis that the proportion of votes cast for Candidate A is 1/2. We cannot accept this hypothesis without counting all the votes. We may, however, find sufficient reason to reject the idea that Candidate A will get 1/2 the vote. If we were to select a well-designed random sample of precincts and find that 60%, 70%, or even 75% of the ballots in the sample were cast for Candidate A, we might be inclined to reject the null hypothesis. Instead of feeling that Candidate A will get 1/2 the vote, we suspect he will get more than 1/2 the vote and win the election. If Candidate A were to get only 40%, 30%, or as little as 25% of the votes in our sample, we would also be inclined to reject the null hypothesis. In this case we would feel that Candidate A is likely to get less than 1/2 the vote or lose the election. If, on the other hand, the proportion of votes cast for Candidate A in our sample is about 1/2, we would not have sufficient evidence to reject the null hypothesis: we would fail to reject the idea that Candidate A will get 1/2 the vote. (In the latter instance we may have to wait until all the votes are counted and we find out the true proportion of votes cast for Candidate A.)

**Deciding How to Set Up the Test**   Setting up a test of the statistical hypothesis that 1/2 the votes will be for Candidate A requires that we decide when a sample proportion is about 1/2. Likewise, we must decide when the sample proportion is far enough from 1/2 to cause us to reject the null hypothesis.

When deciding how to set up a test of hypothesis we must consider the possibilities of making an incorrect decision. (This was discussed in the section on decision making in chapter 4.) Our null hypothesis might be true or false, and we might reject or fail to reject it. Table 1 summarizes 4 decision-making possibilities relative to the null hypothesis. (Recall that our null hypothesis is that the proportion voting for Candidate A is about 1/2. Therefore, if the null hypothesis is true, the election is in fact very close. The errors indicated in the box in table 1 correspond to the Type-I and Type-II errors, which we discussed in chapter 4.)

The consequences of making these two types of error must be considered when setting up a test. A Type-I error results from rejecting

TABLE 1    Type-I and Type-II Errors

| | Null Hypothesis Is | |
| --- | --- | --- |
| | True | False |
| Reject Null Hypothesis | Type-I error | Decision correct |
| Fail to Reject Null Hypothesis | Decision correct | Type-II error |

a true null hypothesis. If the null hypothesis is true, the election can be thought of as one that was in fact too close to call. If the null hypothesis is, on the other hand, false, the public would expect that the networks successfully project a winner. In our example, then, a Type-I error means that the television networks incorrectly project a winner when the election is too close to call. In our last example, ABC and NBC made such an error. A Type-II error would occur if we failed to accurately project a winner when we should have been able to do so.

In practice, we *set* (determine before hand) the probability of making a Type-I error. This probability is called the level of significance for the test.

> The *level of significance* for a test of hypothesis is the probability of committing a Type-I error.

Commonly, the chances of making a Type-I error are set at .1, .05, or .01. For a particular sample size, as the level of significance increases, the chance of making a Type-II error is reduced, and vice versa. Hence, we would choose a high level of significance, let's say .1, if we judged a Type II error to be more serious than a Type-I error. If a Type-I error is considered to be more serious, .01 is used. The compromise level is then .05.

**Setting Up the Test**    In our example, the television executive must decide which is worse — incorrectly projecting a winner (a Type-I error) or not projecting when one of the other networks might correctly project a winner (a Type-II error). Let us set the level for our test at 5% (level of significance is .05) since it is unlikely that the three executives will agree on which type of error is worse and 5% is a good compromise value for illustrative purposes.

Recall that the level of significance is the probability of rejecting a true null hypothesis. In our case 5% is the chance of rejecting the null hypothesis that 1/2 of the votes will be for Candidate A or that the election is too close to call. Suppose we are taking a sample of 600 people. The sampling distribution of the proportion of voters in our sample voting for Candidate A will look like figure 2 if exactly 1/2 of the voters vote for Candidate A. The shaded areas of the graph in figure 2 are the rejection, or critical, region for our test.

> The *rejection*, or *critical*, *region* for a test of a hypothesis represents the values of the statistic that will lead to rejection of the null hypothesis.

If the sample proportion is quite a bit larger than 1/2, greater than **C** on the graph, we will reject the null hypothesis and project A to be the winner. If, on the other hand, the proportion in our sample who voted for A is small, less than **D** on the graph, we will reject the null hypothesis and project A's opponent to be the winner. Hence, if the sample proportion is far away from 1/2, below **D** or above **C**, we will project a winner. If the sample proportion is about 1/2, between **C** and **D**, we fail to reject the null hypothesis and declare the election too close to call.

Since the level of significance (the probability of rejecting a true null hypothesis) has been set at 5%, the shaded area of figure 2 must be the region beyond 2 standard deviations of 1/2 — a 5% area. Above the point **C** in figure 2 the area is 2.5%; the area below **D** is also 2.5%. The total area of the rejection region is

2.5% + 2.5% = 5%

For any normal distribution 95% of the values are within 2 standard deviations of the middle and, hence, 5% of the values are

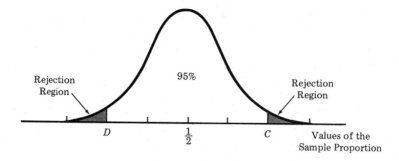

**FIGURE 2**   **Distribution of the Sample Proportion if Candidate A Gets 1/2 the Vote**

more than 2 standard deviations from the middle. (Recall that figure 2 represents the sampling distribution of the sample proportion when the null hypothesis is true. If the null hypothesis is true, the distribution of the sample proportion is nearly normal with mean 1/2. The sampling distribution must be specified when the null hypothesis is assumed true. This is why the null hypothesis involves a statement that the proportion *equals* some value.)

$$\text{standard error of the proportion} = \frac{.5}{24.5} = .02 = 2\%$$

$$C = 50\% + 2(2\%) = 50\% + 4\% = 54\%$$
$$D = 50\% - 2(2\%) = 50\% - 4\% = 46\%$$

(Note that 24.5 is the square root of 600 [Appendix B, table 4]. The standard error of the proportion is the standard deviation of the population, .5, [see Appendix C] divided by this square root.)

Observe that we are working with the standard error of the proportion. We assume the null hypothesis is true when we set up a test. Under this assumption the standard error of the proportion is known — it equals 2% in this example; hence, there is no need to estimate it.

Our 5% test of hypothesis is then:

1. If the sample proportion is in the critical region, reject the null hypothesis and project a winner. Specifically, if the sample proportion is less than $D = 46\%$, we project A's opponent to be the winner; if the sample proportion is greater than $C = 54\%$, we project A to be the winner.

2. If the sample proportion is between 46% and 54% (not in the critical region), do not project — the election is too close to call.

Note that an estimate of the winning proportion will also be made. On election night a television executive first decides to "project" or "not project" (the part of the evening's work which the public is not likely to forget) and then estimates the winning proportion. (Few people will pay attention to the accuracy of this estimate.) The executive is using the inferential processes of estimation and testing.

In the example of the presidential primary in Wisconsin, ABC and NBC took a chance and projected a winner in an election that was too close to call. They lost their gamble, while CBS held to a too-close-to-call position. It soon became apparent that ABC and NBC were wrong. You'll be asked to discuss how these networks might arrive at such different decisions.

## Test of a Hypothesis: Two Populations

A common type of statistical test involves more than one population. We are sometimes interested in whether two populations are the same or different. In the case of two dichotomous populations we would ask whether the proportion of 1s in the two populations is the same or different. Advertising claims that a product is significantly better than a competing product might involve such a test. The two dichotomous populations could be people who might use the two products in question. A user is "satisfied" (1) or "not satisfied" (0). A random sample of people would be randomly placed into two groups — each group would use one of the two products. A company would then ask, "Are the populations that represent the use of these two products the same — that is, are the proportions of 'satisfied' people the same in the two dichotomous populations?" If not, we feel that one proportion is larger than the other, indicating more satisfaction with one product than the other.

In Exercise 18 we will look at the saccharin example we mentioned at the beginning of this chapter. Another example involves a look at the results of two surveys. Recall in chapter 1 and chapter 2 our discussion of statistical doublespeak as the result of conflicting survey results. Two surveys may both be targeted to, for example, adults 18 years of age or older; but, because the surveys were run at different times, used different methods of contacting people, or had different response rates, the two populations actually sampled for the surveys may differ. We might ask whether the results of two surveys are really different, that is, do the survey results differ significantly?

**Null Hypothesis**   Our null hypothesis is that the proportions of 1s in the two populations is equal, or, equivalently, that the difference between the population proportions is zero.

The inferential statistic of interest is the difference between the sample proportions obtained by the two surveys. The difference between the sample proportions has a distribution that is approximately normal, centered about the difference of the population proportions. Figure 3 illustrates this distribution. The difference of the population proportions is zero under the null hypothesis that the population proportions are equal.

**Standard Error**   The standard error for the difference between the sample proportions is not known when the null hypothesis is true. An estimate of this unknown standard error is the square root of the sum of the estimates of the *variances* of the distributions of the two sample proportions.

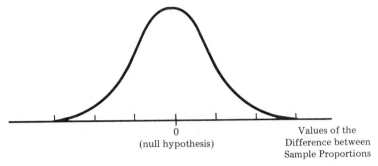

FIGURE 3   Distribution of the Difference Between Sample Proportions (under the null hypothesis that the two
dichotomous populations are the same)

estimated standard error of the difference between proportions
   = square root [sum of estimated variances of proportions]
   = square root [(standard error of one proportion) squared
      + (standard error of second proportion) squared]

It might seem strange that in order to get the estimated standard error
of the difference between two proportions, we first square the estimated
standard errors of the proportions. After adding these squared quantities, we then take the square root. This process must be followed. We
*cannot* add the estimated standard errors. The estimated variances
(square of the estimated standard error) do add.

(The variance of the difference between independent statistics is
the sum of the variances of these statistics. The two sample proportions
in our example are independent since the samples were independently
drawn. The estimated variance of the distribution of the sample proportion is the square of the estimated standard error of the proportion,
which was discussed at the beginning of this chapter.)

**Example**   First, observe that the estimated standard error of the difference between proportions is *not* equal to the sum of the estimated
standard errors of the two proportions. Suppose the estimated standard
errors of the proportions are 3 and 4, a sum of 7. The estimated standard error of the difference is, however,

square root (3 squared + 4 squared) = square root (9 + 16)
                                     = square root (25) = 5

The sum of the estimated standard errors does not equal the estimated
standard error of the difference.

Let us look at the example given in Exercise 3 of chapter 2. In
that example 38% of the Americans surveyed (15 to 17 May 1974) felt

President Nixon should remain in office. (This was during the Watergate scandal, which led to Nixon's resignation in August 1974.) The survey was a telephone poll of 778 people conducted for *Time* magazine by the Yankelovich survey organization.

During that same time a Harris survey found 41% wanted Nixon to remain in office. The Harris poll, a personal interview poll, was probably based on about 1,200 interviews.

The estimated standard error of the difference of the sample proportions in this case would be 2.2%.

$$\text{estimated standard error for Yankelovich estimate of 38\%} = \frac{.49}{27.9} = .018\ (1.8\%)$$

$$\text{estimated standard error for Harris estimate of 41\%} = \frac{.49}{34.6} = .014\ (1.4\%)$$

estimated standard error of difference between percentages
= square root (.018 squared + .014 squared)
= square root (.0003 + .0002)
= square root (.0005) = .022 (= 2.2%)

Figure 4 shows the distribution of the difference between the Yankelovich and Harris sample proportions.

A 5% level of significance test of the hypothesis that the two dichotomous populations (adults with access to listed telephones and all adults) are the same as far as their opinions on this question are concerned would be:

1. If the difference of the sample proportions is in the critical region (less than −4.4% or greater that 4.4%), we reject the null hypothesis that the two dichotomous populations in question are the same. Specifically, if the difference of the sample proportions exceeds 4.4% (2 estimated standard errors above zero), we reject the hypothesis that the two dichotomous populations are the

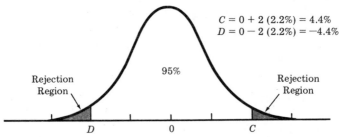

$C = 0 + 2\ (2.2\%) = 4.4\%$
$D = 0 - 2\ (2.2\%) = -4.4\%$

95%

Rejection Region

Rejection Region

$D$      0      $C$

**FIGURE 4**  Distribution of the Difference between Yankelovich and Harris Sample Proportions

same and conclude that the Yankelovich survey found a greater proportion of people wanting Nixon to remain in office. (Note that our subtraction is the Yankelovich proportion minus the Harris proportion.) If, on the other hand, the difference of the sample proportions is less than $-4.4\%$, we reject the null hypothesis, concluding that the Harris poll found a greater proportion.

(A test is set up before the data are collected. Prior to seeing the sample results we could not rule out the possibility that the Yankelovich survey might find a greater percentage than the Harris survey. This is why both alternatives must be considered.)

2. If the difference between the sample proportions is between $-4.4\%$ and $4.4\%$ (not in the critical region), we fail to reject the hypothesis that the two dichotomous populations are the same.

Since the results of our samples give a difference of

$$38\% - 41\% = -3\%$$

our results are as described in action 2. This apparent "doublespeak" is not an unusual difference at all but, rather, a difference we can expect when selecting random samples. (We'll look at the third survey conducted on the question of Nixon remaining in office in Exercise 2 of this chapter.)

## Estimating a Population Mean

We talked about estimating the proportion of 1s in a dichotomous population in the first section of this chapter. The proportion of 1s in a population of 0s and 1s is the mean of that population. An extension of that discussion would involve estimation of the mean of a population that is a collection of continuous numbers, not a collection of just 0s and 1s. For example, a researcher might be interested in the mean weight-gain of hogs fed a certain ration. Suppose the researcher fed a simple random sample of 50 hogs an experimental ration and found a mean weight-gain of 8.8 lb. This statistic would serve as a point (single-number) estimate of the mean gain for this breed of hogs when fed this ration.

How accurate is such an estimate? From our discussions in chapter 5 on the sampling distribution of the mean we recall that the sample mean, as the sample proportion, varies from simple random sample to simple random sample in a normal pattern. It follows, as with the sample proportion, that the sample mean is within two standard errors of the population mean with 95% probability. Or, the population mean is within two standard errors of the sample mean for 95% of the random samples we might draw.

Since the standard error of the mean is not necessarily known, it is estimated. The estimated standard error of the sample mean is the standard deviation of the observations in the sample (the sample standard deviation) divided by the square root of the sample size.

$$\text{estimated standard error of the mean} = \frac{\text{standard deviation of sample}}{\text{square root of the sample size}}$$

For 95% of the samples we might select, the population mean will be within 2 estimated standard errors of the sample mean. (Again, we assume the samples are large so that the normal model is appropriate.) In the feed example, suppose that the 50 sample weight-gains had a standard deviation of 3 lb. (See the section "Descriptive Statistics: The Mechanics" in chapter 3 for a discussion of how you might find this sample standard deviation or the standard deviation of the 50 numbers in the sample.) The estimated standard error for our sample mean is

$$\frac{3}{7} = .42$$

(The square root of 50 is about 7. See Appendix B, table 4.) A 95% confidence interval for the mean weight-gain for this breed of hog is then 7.96 lb. to 9.64 lb.

$$8.8 + 2(.42) = 8.8 + .84 = 9.64$$
$$8.8 - 2(.42) = 8.8 - .84 = 7.96$$

## Tests of Hypotheses Concerning the Population Mean

We present here examples of three types of tests of hypotheses about the mean of a continuous population. A statistical hypothesis can be thought of as a statement about a parameter of a population. In this section we will consider the population mean as the parameter of interest.

### Mean Equals Specified Value

Let us first look at the test that the mean of a population of continuous data equals a specified value. Suppose that an electronics firm that produces light bulbs claims in its advertising that the mean life of its bulbs is 800 hours. This company continually checks samples of its bulbs to make sure that its claim is accurate. For example, let's say a simple random sample of 39 bulbs is tested and the hours until the bulbs burn out listed as shown here.

833, 704, 802, 738, 700, 641, 822, 818, 830, 820, 705,
622, 691, 820, 904, 830, 820, 740, 800, 921, 607, 633,
628, 720, 804, 738, 827, 683, 721, 777, 809, 639, 747,
808, 881, 910, 707, 697, 811

If the (population) mean life is in fact 800 hours (our null hypothesis), we know from the section "Sampling Distribution of the Mean" in chapter 5 that the sample mean will vary normally about 800 as shown in figure 5. As discussed in the previous section of this chapter, the standard error of the sample mean must be estimated. The standard deviation of the 39 numbers in our sample is 776, so the estimated standard error is

$$\frac{776}{6.2} = 125.2$$

We therefore expect the sample mean to be within 2 estimated standard errors of 800, or between 552 and 1,048, if the null hypothesis is true.

800 + 2(124) = 800 + 248 = 1,048
800 − 2(124) = 800 − 248 = 552

Our 5% test becomes:

1. If the sample mean is in the critical region (less than 552 or greater than 1,048), reject the null hypothesis that the (population) mean life of the bulbs is 800 hours. Specifically, if the sample mean is less than 552, reject our null hypothesis and conclude that the (population) mean life is less than 800. If the sample mean exceeds 1,048, reject our hypothesis in favor of the conclusion that the (population) mean life is more than 800 hours.

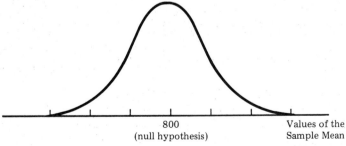

800
(null hypothesis)

Values of the
Sample Mean

**FIGURE 5**   **Distribution of Sample Mean if Population Mean Is 800**

2. If the sample mean is between 552 and 1,048, there is no evidence to contradict the company's advertising claim.

Since the sample mean is about 762, we fail to reject the null hypothesis, so there is no reason to doubt the advertising claim of the company.

### Two Independent Samples

Our next two examples involve the comparison of the means of two populations. First, recalling our example in the previous section of this chapter, suppose we have two rations for hogs that we would like to compare. Let two independent simple random samples of 50 hogs be selected. Each sample of hogs is fed one of the two rations. The mean gains for each ration are recorded for each sample of hogs.

The difference between the sample means will vary normally about zero under the hypothesis that the population means are equal. (Recall the analogous statement concerning sample proportions in the section on testing a statistical hypothesis earlier in this chapter.) The estimated standard deviation of this normal distribution, the estimated standard error of the difference between the sample means, is the square root of the sum of the estimated variances of the distributions of the two sample means. If the sample variances are 9 and 25, respectively, the estimated variances of the sampling distributions of the two means are 9/50 and 25/50, respectively. The formulas for the estimated variances of the sampling distribution of the mean are

$$\begin{array}{c}\text{estimated variance} \\ \text{of the sampling distribution} \\ \text{of the mean}\end{array} = \frac{\text{sample variance}}{\text{sample size}}$$

The estimated standard error of the difference between sample means is

$$\begin{array}{c}\text{estimated standard error} \\ \text{of difference} \\ \text{between samples means}\end{array} = \begin{array}{c}\text{square root (sum of estimated} \\ \text{variances of the two means)}\end{array}$$

The estimated standard error of the difference between the sample means is then the

$$\text{square root}\left(\frac{9}{50} + \frac{25}{50}\right) = \text{the square root } (.68) = .8$$

Hence, if the two population means are equal, the difference between the two sample means varies normally about zero with an estimated standard error of .8.

Our 5% test becomes:

1. If the difference between the sample means is less than $-1.6$ or greater than 1.6 (in our critical region), reject the hypothesis that the population means are equal and conclude that the mean is largest for the population whose sample mean is largest.

2. If the difference between the sample means is between $-1.6$ and 1.6 (within 2 estimated standard errors of zero), fail to reject the hypothesis that the means of these two populations are equal (the mean gain is the same for both rations).

Suppose in this example that the two sample means are 8.8 and 6.7 giving a difference of 2.1. We therefore reject the null hypothesis and conclude that the better ration is the one given to the sample of hogs whose mean weight-gain was 8.8 lb. (a better ration is one in which the population mean gain is larger).

## Two Samples: Not Independent

The third and final example in this section involves the test that two populations have the same mean when the simple random samples are not independent. In this example an element from one population is *paired* or *matched* to an element from the other population. "Before" and "after" observations are a classic example. Suppose that a test is given to 96 school children before and after a special class. The change or difference in a child's score indicates whether the education program was beneficial to that child. The two populations of interest here are a set of students before receiving this special class and a set of students who received the special training.

A student's score on an exam before the special class is paired with his or her score on an exam after the special class. In such a case the difference between a student's test scores is of interest (the score after the special class is subtracted from the test score obtained before the class).

difference = score before − score after

This difference is an indication of the change that resulted from the special class. The sample of differences represents a *single sample in a population of differences* and is treated as the single population example discussed previously. If the special class has an effect, the mean of the population of differences is less than zero. The greater-than-zero alternative is not considered here since we assume that this special course will not "hurt" students.

Observe that figure 6 graphically illustrates the distribution of the sample of differences between "before" and "after" test scores. Figure 7 shows information from Appendix B, table 1 that is pertinent to

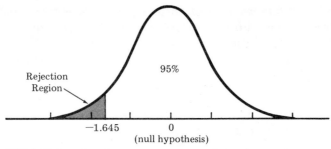

**FIGURE 6    Graphic Representation of Test-Scores Test**

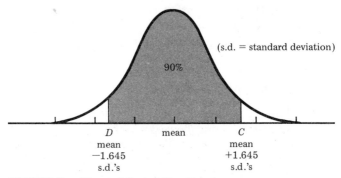

**FIGURE 7    Normal Probability Graph**

our example. Note that for our sample of test score differences the total area above point **D** is .95.

The estimated standard error of the mean of the sample of differences is

$$\text{square root} \left( \frac{4.89}{96} \right) = .23$$

(4.89 is the variance of the sample of differences.) Then,

$$0 - 1.645(.23) = -.38$$

The 5% test is therefore:

1.  Reject the null hypothesis if the mean of the sample differences is less than −.38 (in the critical region) and conclude that the special course does improve student performance as measured by this test.

2.  Fail to reject the null hypothesis if the sample mean is greater than −.38.

Suppose that the mean of the sample of differences is −.82. The null hypothesis is therefore rejected. We conclude that the special training had a beneficial effect.

# Nonparametric Statistics

The statistical inference we have discussed to this point can be thought of as large sample theory — that is, we have been assuming that the size of the sample drawn from the population of interest is large enough that we can assume that the sampling distribution of the mean is normal. (Recall our discussions on the distribution of the sample proportion and the sampling distribution of the mean in chapter 5.) For the statistical inferences thus far discussed, the normal model is appropriate only for large samples. But sample sizes are not always large enough. Maybe only a few people or only a few laboratory animals can be tested. Is the normal model we have discussed still appropriate? Not necessarily. For small sample sizes the sampling distribution of the proportion, or, more generally, the sampling distribution of the mean, may not be normal. Thus, when sample sizes are small, our normal model may not be appropriate even when the original population is normal.

In chapter 5 we stated that the sampling distribution of the proportion and mean are normal if the sample size is large. We said little about the distribution of the numbers in the original population. We looked at both symmetric and skewed populations when discussing the Central Limit Theorem. Larger sample sizes are required for the sampling distribution of the mean to be normal when the original population is skewed. The normal model is not adequate when samples are small, even if the distribution of the data in the original population is normal. Student's $t$-distribution would actually be the basis for the probability model that would be used. This distribution will not be discussed.

We will not be discussing small sample theory, but we need to be aware that our normal model may not be appropriate if sample sizes are not large. Should you study statistics further you will indeed become involved with the problem of small sample sizes. Our interest in this section is with looking at the distribution of numbers in a population in a nonparametric manner.

When the parameters, such as the mean and standard deviation in the normal case, are used to describe the distribution of data in a population, statistical inference is said to be ***parametric***. For a more general description of the distribution, *nonparametric* or *distribution-free* statistical inference results. A more general description might be to say that the data in a population have a symmetric distribution. This is a more general description than the normal parametric form as there are many nonnormal distributions that are symmetric.

We will look at two types of nonparametric tests in the remainder of this section. We will consider a *sign test* and two Wilcoxon *rank tests*. These tests do assume a normal model, but for sample sizes that may not be as large as previously assumed.

### Sign Test

Let us consider testing a hypothesis concerning the median of a population. (For symmetric populations the mean and median coincide so in that case the test we present here is also appropriate for testing the mean for a population.) The word "sign," which is used to describe the test to be discussed in this section, refers to whether numbers are positive or negative. We therefore get back to a dichotomous situation. Let us look at a one-population example.

Our example involves a test that the median of a population is a specified value. We will be considering a random sample of 20 faculty members from the college of education at a university. The median income in the United States for the year that the data were collected was $11,116. We wonder if the faculty of the college of education earns more than the average income for United States families.

Observe that if the median salary of the faculty is indeed $11,116, then a faculty member selected at random will have a 50-50 chance of having a salary about $11,116. Equivalent to saying the median salary is $11,116 is to say that 1/2 the faculty have salaries above $11,116. We therefore dichotomize our data by representing salaries about $11,116 by a 1 and all other salaries by a 0. (Think of this as looking at the differences of salaries from $11,116. Positive differences are represented by a 1, and negative differences are represented by a 0. For example, a salary of $12,150 has a deviation of $12,150 − $11,116 = $1,034, which is represented by a 1, and a salary of $9,425 has a deviation of $9,425 − $11,116 = -$1,691, which is represented by a 0.) Equivalent to the hypothesis that the median salary of the population is $11,116 is the hypothesis that the proportion of salaries above $11,116 is 1/2. We are therefore back to the case of testing that the proportion of 1s in a population is 1/2. (This specific test is to be completed in Exercise 7.)

Observe that our analysis of these data is not direct because we did not use the statistic that corresponds to the parameter of interest — that is, we did not use the sample median for our inference to the population median. Also observe that under the null hypothesis the sampling distribution of the sample proportion should be close to normal for rather small sample sizes.

### The Wilcoxon Signed-Rank Test

Recall from our discussion of the sign test that we were only concerned with whether the data exceeded a certain value. The sign test ignores the size of the differences. In our present discussions we will look at the test proposed by Frank Wilcoxon in the mid-1940s. The Wilcoxon tests will take into account the magnitude of the differences of interest.

Again our interest is with the normal model. We assume in this section that the data in the population of interest is symmetrically but not necessarily normally distributed. Let us look at data similar to the data considered in the previous section of this chapter.

The data consists of the bulb life of 12 light bulbs (in hours). The data are

1158, 701, 1046, 947, 708, 912, 875, 850, 713, 913, 706, 1147

Suppose we are interested in testing whether the mean life of this brand of bulbs is 1,000 hours. We will test this hypothesis by first looking at the deviations of the data from 1,000. These deviations are:

$$1158 - 1000 = 158, 701 - 1000 = -299, 46, -53, -292, -88,$$
$$-125, -150, -287, -87, -292, 147$$

We would then *rank* the deviations ignoring the signs. The largest rank of 12 goes to the value $-299$; $-294$ has a rank of 11; etc. This procedure is summarized in table 2. The *signed ranks,* as given in

TABLE 2    Ranked and Signed Rank Deviations for Bulb Life Data

| Bulb Life | Deviations from 1000 | Ranks | Signed Rank |
|---|---|---|---|
| 1046 | 46 | 1 | 1 |
| 947 | $-53$ | 2 | $-2$ |
| 913 | $-87$ | 3 | $-3$ |
| 912 | $-88$ | 4 | $-4$ |
| 875 | $-125$ | 5 | $-5$ |
| 1147 | 147 | 6 | 6 |
| 850 | $-150$ | 7 | $-7$ |
| 1158 | 158 | 8 | 8 |
| 713 | $-287$ | 9 | $-9$ |
| 708 | $-292$ | 10 | $-10$ |
| 706 | $-294$ | 11 | $-11$ |
| 701 | $-299$ | 12 | $-12$ |

table 2, are the ranks of the deviations along with the sign, plus (+) or minus (−), of the corresponding deviations.

The Wilcoxon signed rank statistic of interest is the sum of the positive ranks. From table 2 we see that this value is $8 + 1 + 6 = 15$. Observe that if this number is large, the data tend to deviate above 1,000 — that is, the mean life of the bulbs is greater than 1,000. On the other hand, a small value for the Wilcoxon statistic would indicate that the data tends to be less than 1,000.

Our test will be determined using a standardized form of the Wilcoxon statistic. To standardize a statistic we subtract the mean of the sampling distribution of the statistic from the statistic. We then divide this difference by the standard error of the statistic. The standardized Wilcoxon statistic has an approximate standard normal distribution. Recall from our discussion that standard normal data has a mean of 0 and a standard deviation of 1. To standardize a statistic we adjust it so that the mean is 0 and the standard error is 1. We accomplish this by the following adjustment

$$\text{standardized statistic} = \frac{\text{statistic} - \text{mean of statistic}}{\text{standard error of statistic}}$$

A 5% test based on a standard normal statistic would be to:

1. Reject the null hypothesis if the statistic is less than −2 or greater than 2.

2. Fail to reject the null hypothesis if the standard normal statistic is between −2 and 2.

Appendix B, table 5 gives us the necessary standardizing values. From table 5 we see that for a sample size of 12 the mean and standard error of the Wilcoxon signed-rank statistic are 39.0 and 12.7, respectively. Standardizing the value of the statistic, which is 15, we get

$$(15 - 39)/12.7 = -1.9$$

We therefore fail to reject the hypothesis that the mean bulb life is 1,000 hours.

### Wilcoxon Rank-Sum Test

In 1945 a test that two populations are the same was also provided by Wilcoxon. Let us look at an example. Suppose we are testing whether a new feed ration for cattle will increase the weight gain of the animals above the gain in weight that comes from the usual feed ration. Let us say the weight gains for 8 cattle fed the new ration are

54, 58, 63, 68, 72, 75, 84, 109

The 6 cattle fed the usual ration have weight gains of

69, 77, 85, 98, 114, 134

The two samples are combined and the values ranked from 1 to 14 as shown in table 3.

We will let the Wilcoxon (two-sample) rank-sum statistic equal the sum of the ranks for the sample with the most observations. (If both samples are the same size, it makes no difference which is used.) In our example we would sum the ranks of the sample of cattle that received the new ration as there are more observations in that group than there are for the sample of cattle receiving the usual ration. This sum is

$$1 + 2 + 3 + 4 + 6 + 7 + 9 + 12 = 44$$

Observe that the larger this value becomes, the more inclined we are to think that the new ration is superior to the usual ration. On

TABLE 3  Ranking for the Wilcoxon Two-Sample (Rank-Sum) Statistic

| New Ration | Rank | Usual Ration | Rank |
|---|---|---|---|
| 54 | 1 | | |
| 58 | 2 | | |
| 63 | 3 | | |
| 68 | 4 | | |
| | | 69 | 5 |
| 72 | 6 | | |
| 75 | 7 | | |
| | | 77 | 8 |
| 84 | 9 | | |
| | | 85 | 10 |
| | | 98 | 11 |
| 109 | 12 | | |
| | | 114 | 13 |
| | | 134 | 14 |

the other hand, a small value for the Wilcoxon statistic indicates that the new ration is not as good as the usual ration.

Just as with the Wilcoxon one-sample test statistic, we standardize the Wilcoxon two-sample statistic. The standardizing values, given in Appendix B, table 6, are the mean and standard error (deviation) of the Wilcoxon rank-sum statistic. From table 6 we see that the mean of the sampling distribution of our statistic is 60, and the standard error is 7.7. (Note that these values are found by looking in table 6 for the values corresponding to a larger sample size of 8 and a smaller sample size of 6.) Our standardized statistic is found, as before, by subtracting the mean and then dividing by the standard deviation. This value is

$$(44 - 60)/7.7 = - 2.1$$

Our test is as given in the one-sample example; that is, reject the null hypothesis that the mean weight gain for the two rations is the **same** if the standardized statistic exceeds 2 or is less than $-2$. We therefore reject the null hypothesis and conclude the new ration does not even do as well as the usual ration.

## The Role of Descriptive Statistics in an Inferential World

It would be desirable at this time to stop and reinvestigate the difference between descriptive and inferential statistics.

Descriptive statistics are numbers used to describe a collection of numbers we called a data set. The data set that is to be described could be the entire population of interest. For example, FBI crime statistics are an attempt to describe crime as it affects the entire population. (See Exercise 1, chapter 2.) There is no sampling of the police agencies that furnish data to the FBI. Noncoverage comes from nonresponse; that is, the only crimes not included in FBI statistics are the crimes that the victims do not report to the police or the crimes not reported by the police agencies to the FBI. Every attempt is made to describe all crimes in the country.

In chapter 3 we described certain other complete populations. Recall figure 2, chapter 3 graphically illustrated the size of all households, the education of all adults, and the size of all families in the United States. Descriptive statistics can, therefore, be used to describe the entire data set of interest.

As mentioned in both chapter 1 and chapter 4, there are instances when the population of interest cannot be studied in its entirety. Only a part of the population can be investigated. We may then use sample data to infer to different aspects of the population. Inference is not necessarily restricted to the mean of a population. Recall that the distribution, central tendency (mean, median, and modes), and dispersion (range, standard deviation, and percentiles) are all likely to be important characteristics of a population. Inference to any of these descriptive characteristics of the population is possible.

To get an overview of an unknown population, we might describe the sample in some detail. It is quite possible that we would like a description of the population under investigation. When the population cannot be studied, a description of the sample drawn from the population is the next-best thing. As a matter of course, you should look at the sample in some detail before getting involved with inference to population characteristics. You could, therefore, be using a statistic to describe the sample and then using the same statistic to infer to a population characteristic — that is, a statistic may be both descriptive and inferential. Recall, however, that when a statistic becomes inferential, uncertainty enters the picture.

A word of caution is in order. A description of a sample is not justification for inference. Let us look at an example. A sample may not have been randomly selected. Showing that the sample is "like" the population of interest does not justify inference to the population. We discussed this in the section "Other Sampling Procedures" in chapter 5. Recall the survey of readers of a magazine. The 100,000 respondents to this voluntary response survey were described according to a number of characteristics such as education, marital status, and age. These sample statistics were about the same as the corresponding figures on all females in the country. (The Census Bureau publishes such information.) The fact that the sample appeared to be "like" the population was used to justify the claim that the results of the survey represented the views of all United States females.

Inference in such cases is very questionable. Recall that the normal model gives a measure of uncertainty. For the normal model to be appropriate random selection of sample items is required. Without random selection the uncertainty involved with inferring from the sample to the population cannot be measured. Such inference is therefore meaningless as the relationship between the unknown population parameters and the corresponding sample statistics is not known.

When a population of interest cannot be investigated, a description of the sample that is the basis for inference can be helpful. Note from our discussions here and in chapter 5 that a description of a sample may either signal that the sample is not a good sample or give an adequate description of an unknown population.

## Summary

This chapter has provided a look at how statistical inferences are made. For a dichotomous population we said that inference is restricted to the proportion of 1s in the population. This follows since the proportion of 1s in a dichotomous population (a parameter) completely determines the population. The proportion of 1s in a sample serves as an approximation of this unknown parameter. For random samples the error of this estimation procedure can be predicted.

A confidence interval for the proportion of 1s in a dichotomous population takes into account the variation of the sample proportion from random sample to random sample. A 95% confidence interval was found to be a two-estimated-standard-error interval about the sample proportion. The process of taking random samples and finding the corresponding 95% confidence intervals will result in intervals having the unknown population proportion within their limits for 95% of the samples that are selected; 5% of such intervals will not contain this parameter.

A test of hypothesis about the proportion of 1s in a dichotomous population was discussed. A determination of the level of significance specifies a critical region. If our sample proportion falls in this region, the hypothesis being tested, the null hypothesis, is rejected. (Even though prior feelings and feelings about the consequences of making errors [related to Type-I and Type-II errors] were not formally brought into the analysis, these tests of hypothesis were seen to be decision problems, which are influenced by these factors. Recall our discussion in chapter 4 of the decision-making process.)

In addition we mentioned tests involving two dichotomous populations. We said, for example, that two dichotomous populations could be the basis of advertising claims of significance. "Equal" populations may be translated into "equivalent" products; the alternative to equality implies the superiority of one product over another.

The saccharin experiment mentioned at the beginning of this chapter is another example. The test rats, rats fed saccharin, and the control rats, rats not fed saccharin, represent samples of two populations. (We will discuss the design of such an experiment in chapter 8.) A test that the proportion of rats developing bladder tumors was the same for these two groups of animals was the basis for the partial ban of saccharin in the United States. (See Exercise 18.)

A third example of testing whether two dichotomous populations are the same involved us with seemingly contradictory survey results. As discussed in chapter 1 and chapter 2, statistical doublespeak can occur when surveys show contradictory results. Because of different background influences on surveys, the populations actually sampled by two survey groups may be different. We can run a test to see if these populations do indeed differ. Knowing whether the populations are different can help us avoid doublespeak.

Statistical inference concerning the mean of a population was discussed in two parts in this chapter. We first looked at parametric inference based on large samples. For large samples the sampling distribution of the mean is approximately normal. The near normality of this distribution made our use of the normal model an adequate way of measuring the uncertainty of statistical inference. If sample sizes are not as large and the data in the population has a symmetric distribution, the normal model can still be adequate through the use of sign tests or Wilcoxon (rank) tests.

We have seen how the few normal probabilities summarized in Appendix B, table 1 give us a model allowing us to measure the uncertainty in statistical inference. This measurement is crude as only a few probabilities are discussed (68%, 95%, and 99.7% being the primary probabilities of interest). Do not get the false impression that the normal distribution can give us probability measures for the uncertainty of all types of statistical inference. This is not so. The chi-square, the Student's *t,* the *F*-distribution, and the binomial distribution all provide measurements of the uncertainty of different statistical inferences, that is to say, these distributions model different sampling situations. Hopefully, you have developed an understanding for interpreting inferential statistics without being overwhelmed by so many different probability distributions. Further studies into statistical inference will necessitate a study of other probability distributions. The basic idea of inference will not change however. (In chapter 9 we will look at a chi-square model.)

One more thing before we close our present discussions. The statistical errors, that is, the probability models discussed in this chapter, are based on the selection of a random sample (simple random samples). If a sample is not randomly selected, the model may not be adequate. Indeed, if the background influences discussed in chapter 2 do have an adverse effect on the data, the model used in this chapter could do a very poor job of describing the actual sampling process. Do not forget that our interpretation of inferential statistics must take into account the background influences we discussed earlier.

# No
# Comment

■   *"Last week the Federal Government's Center for Disease Control announced that a certain drug company may have infected 5,000 hospital patients with contaminated intravenous solutions, contribut-*

*ing to the deaths of 500 people. When asked how this figure had been determined, a Government spokesman said that one estimate of 2,000 was 'unrealistic' and another estimate of 8,000 was 'unfair.' So the authorities split the difference."*

(Time, *2 August 1971*)[2]

# Exercises

1.  Referring to Appendix B, table 1, how would you construct 99% or 90% confidence intervals for the proportion of 1s in a dichotomous population or the mean of a continuous population? Use the data of the sections "Estimating a Population Proportion" and "Estimating a Population Mean" in this chapter to work out some specific examples. How would such intervals be interpreted?

2.  In Exercise 3 of chapter 2 contradictory findings of three surveys were presented. We looked at two of these surveys when we discussed testing a hypothesis relative to two populations in this chapter. Compare the results of the Harris and White House polls. Recall that the Harris poll reported that 41% of about 1,200 adults surveyed thought Richard Nixon should remain in office. The White House poll reported 55% of the 1,677 surveyed felt that way. If doublespeak exists here, what might be its cause? (Recall our chapter 2 discussions of the background influences on survey statistics.)

3.  Table 4 concerns the Gallup survey mentioned in the section "Other Sampling Procedures" in chapter 5. Are the results listed in table 4 "significant?" What does significance mean in this instance?

**TABLE 4    Results of Gallup Survey**

| Brand of Color TV Most Recently Purchased | Would Buy Again | Would Not Buy Again | Number of Interviews |
|---|---|---|---|
| Zenith | 82% | 28% | 2118 |
| Brand X | 70% | 30% | 2140 |

4. Before walking to work in the morning, I make an informal test of the hypothesis "It will rain today." Depending on the results of this test, I decide to carry or not to carry an umbrella. What are the consequences of making a Type-I or Type-II error?

5. A company must continually test the level of different chemicals in drugs they manufacture. Suppose that the concentration of a chemical will make a drug "lethal" or "safe." What are the consequences of making a Type-I or Type-II error in this instance? Should the level of significance be large or small?

6. In chapter 4 we discussed the gambler's fallacy. As an illustrative example, we asked how one would bet on the eleventh flip of a coin after 10 successive heads. We noted that 10 heads in 10 flips suggests that the chance of flipping a head is more than 1/2. Test the hypothesis that the coin is fair given that 10 flips resulted in 10 heads. Do you agree that the chances of heads is greater than 1/2?

7. In the section on nonparametric statistics in this chapter we discussed testing that the median of a continuous population was a specified value by dichotomizing the data. Review that discussion, and then test that the median is $11,116 for the population from which were drawn the 20 faculty salaries. (This population is the college of education of a university.) The salaries are as given.

| | | | | |
|---|---|---|---|---|
| $33,000 | $14,500 | $10,660 | $17,500 | $18,350 |
| 10,250 | 22,700 | 14,000 | 8,000 | 7,408 |
| 13,800 | 23,500 | 10,710 | 24,250 | 10,410 |
| 15,060 | 11,500 | 16,250 | 9,900 | 9,778 |

8. Given 15 salaries from an arts and science college of a university, test that the median salary is $11,116.

| | | | | |
|---|---|---|---|---|
| $11,550 | $21,900 | $12,567 | $12,700 | $19,961 |
| 21,850 | 11,650 | 10,000 | 12,100 | 22,900 |
| 20,050 | 21,825 | 10,000 | 12,430 | 12,625 |

9. Run Wilcoxon one- and two-sample tests on the data analyzed in Exercises 7 and 8. Any difference in the conclusion between these tests and the sign tests run previously?

10.   Bennett and Franklin in *Statistical Analysis in Chemistry and the Chemical Industry* give data on two methods (the standard dichronate titrimetric method and a new spectrophotometric method) for determining iron ore content of ore samples.[3] The data is given in table 5. Analyze the data in table 5 using the Wilcoxon signed rank-sum test.

11.   Suppose that a particular machine is used to fill four-ounce packages with an instant pudding mix. A modification in the method to increase the speed of the process is proposed. An experiment is run to see if the new method does indeed properly fill the packages. For the 50 packages filled by the modified method the mean net contents of the packages was 4.075 ounces with a standard deviation of .05. Does the new method properly fill the packages?[4]

12.   Suppose that at the same time the experiment in Exercise 11 was run on the new filling process, the old method filled 50 bags with a mean net contents of 4.091 ounces. (The standard deviation is again .05.) Is there a difference between the two methods?

TABLE 5    Bennett-Franklin Data

| Sample | Standard Method | New Method | Difference |
|--------|-----------------|------------|------------|
| 1  | 28.22 | 28.77 | +.05 |
| 2  | 33.95 | 33.99 | +.04 |
| 3  | 38.25 | 38.20 | −.05 |
| 4  | 42.52 | 42.42 | −.10 |
| 5  | 37.62 | 37.64 | +.02 |
| 6  | 36.84 | 36.85 | +.01 |
| 7  | 36.12 | 36.21 | +.09 |
| 8  | 35.11 | 35.20 | +.09 |
| 9  | 34.45 | 34.40 | −.05 |
| 10 | 52.83 | 52.86 | +.03 |
| 11 | 57.90 | 57.88 | −.02 |
| 12 | 51.52 | 51.52 | .00 |
| 13 | 49.59 | 49.52 | −.07 |
| 14 | 52.20 | 52.19 | −.01 |
| 15 | 54.04 | 53.99 | −.05 |
| 16 | 56.00 | 56.04 | +.04 |
| 17 | 57.62 | 57.65 | +.03 |
| 18 | 34.30 | 34.39 | +.09 |
| 19 | 41.73 | 41.78 | +.05 |
| 20 | 44.44 | 44.44 | .00 |
| 21 | 46.48 | 46.47 | .01 |

13. Members of households in two communities were compared to see if the amount of time spent viewing television was the same for the two communities. The results of a study showed that:

    Community 1: sample size 45, mean = 18.2 hours per week, standard deviation = 6.3 hours

    Community 2: sample size 50, mean = 14.9 hours per week, standard deviation = 5.1 hours

    Do people in these two communities watch television the same amount of time?

14. Suppose that during a certain week the mean weight of products manufactured by a company was 2.62 lb. (The sample was of 200 items. The standard deviation of the weight of these products was .54 lb.) The next week a different lot of raw materials was used. A sample of 190 items gave a mean weight of 2.81 lb with a standard deviation of .49 lb. Had the mean weight of these products increased significantly from one week to the next?

15. The mean nicotine content of 50 samples of a certain make of cigarette was found to be 26.5 grams with a standard deviation of .5 grams. The manufacturer of the cigarette claims that the nicotine level does not exceed 26.4 grams. Should the manufacturer's claim be challenged?

16. Forty-eight first graders were divided into 24 matched pairs. All students would be given an IQ test; but one member of each pair spent 4 twenty-minute periods with the examiner before the test was administered. The differences in the IQ scores between the pairs (treated-control) were $-17$, $-15$, $-8$, $-7$, $-10$, 4, $-10$, 4, 12, 1, 21, 2, $-20$, 2, 0, 5, $-7$, 19, $-8$, 9, $-3$, $-7$, $-1$, and 34. Use the Wilcoxon test to see if familiarity with the examiner improves performance on the test.[5]

17. The number of days until tumor development was recorded for two sets of mice. One set of mice acted as a control, and another set of mice received a treatment from a drug that was to retard tumor development. Table 6 provides the data on how long it took for tumors to develop. Run a Wilcoxon two-sample test to see if the drug is effective.

TABLE 6   Days until Tumor Development for Two Sets of Mice

| Drug | Control |
|------|---------|
| 21 | 4 |
| 14 | 17 |
| 31 | 5 |
| 29 | 6 |
| 33 | 8 |
| 44 | 14 |
| 23 | 3 |
|    | 2 |

18.  In the saccharin experiment that we have discussed 14 of 100 second-generation test rats developed bladder tumors. Information on the 100 control rats has not been made known. For illustrative purposes, suppose two of the 100 control rats developed bladder tumors. Did significantly more test rats develop bladder tumors than did the control rats? Three of 100 first generation test rats developed tumors. Is this significantly different from the control level of 1 in 100?

# For Discussion

1.  On a graph similar to figure 1 in this chapter indicate the values of the sample proportion within two standard deviations of the population proportion. Does the statement make sense that the population proportion will be within two standard errors of the sample proportion with 95% probability?

2.  Referring to the influences on making a decision (see chapter 4), discuss a network executive's decision to "project" or "not project" the winner of an election on election night. What effect might these influences have on the choice of a level of significance for the appropriate test of hypothesis, which we discussed in the section on testing a statistical hypothesis in this chapter?

3.  Earlier in chapter 6 we found that 7.96 lb to 9.74 lb is a 95% confidence interval for the mean weight-gain of a certain breed of hog fed a particular feed ration. Describe the population (of numbers) that is of interest here. Is the mean of this population between 7.96 lb and 9.64 lb? Interpret this interval.

4.  What populations were investigated in the test of hypothesis given in the section "Tests of Hypotheses Concerning the Popu-

lation Mean" in this chapter? Would the populations of *interest* differ from the populations actually investigated?

5.  Cluster sampling was described in the preceding chapter and arose again in our discussion of election-night projecting. In the latter case precincts, or clusters of voters, were selected. Since cluster sampling of this type can have a profound effect on error estimates of proportions, let us look more closely at an example.

    In a survey of an SMSA, 100 blocks or clusters were selected with 5 households selected from each block. This made a total of 500 households. The error for an estimate of 50% was 5%, which equals .5 divided by the square root of 100, not .5 divided by the square root of 500. In other words, clustering increased the error to a level equivalent to a "sample size" equal to the number of clusters selected, rather than for the number of elements selected.

    Clustering, a relatively inexpensive sampling procedure (see the section in chapter 5 on other sampling procedures), increases error to a level mentioned in our example when elements within a cluster are homogeneous. To improve error with clustering, clustered elements should be heterogeneous (at best a "minipopulation"). (In contrast, stratification decreases error when homogeneous elements are put in the same strata.)

    What gain can be expected from clustering when such a high price in error may have to be paid? Consider other clustering examples from chapter 5. Are elements within clusters homogeneous or heterogeneous? What effect might clustering by precincts have on the projection examples of the section on testing a hypothesis relative to two dichotomous populations in this chapter?

6.  The size of a whale population might be estimated in the following way: a sample of whales is "tagged" by shooting a one-foot cylinder into the blubber of the animals; then fishermen are asked to report the number of tagged whales they kill. If, for example, a sample of size 100 whales is tagged and 10% of the whales killed by fishermen have tags, we estimate the population to be of size 1,000 (10% of all whales killed are tagged and a sample of size 100 is tagged — 10% of 1,000 is 100).

    Discuss the different sources of error for such a procedure.[6]

7.  Table 7 shows how people in a Knight–CBS opinion poll responded to the question, "What about other people who have been charged or convicted in Watergate crimes — should President

TABLE 7   Knight–CBS Poll Results

| | Should Pardon | Should Not | No Opinion |
|---|---|---|---|
| Total (N = 629) | 55% | 29% | 16% |
| By Party Preference (%) | | | |
| Democrats | 62 | 26 | 12 |
| Republicans | 45 | 32 | 23 |
| Independents | 52 | 31 | 17 |
| By Sex (%) | | | |
| Males | 57 | 32 | 11 |
| Females | 53 | 26 | 21 |
| By Region (%) | | | |
| East | 53 | 32 | 15 |
| Midwest | 53 | 30 | 17 |
| South | 59 | 26 | 15 |
| West | 53 | 26 | 21 |
| By Education (%) | | | |
| Eighth Grade | 60 | 15 | 25 |
| High School | 58 | 29 | 13 |
| College | 41 | 45 | 14 |

*Source:* Courtesy of Knight-Ridder Newspapers, Inc.

Ford pardon them or not pardon them?" This survey followed the pardon of President Nixon by President Ford for any Watergate-related crimes that Nixon might have committed as president. The figures in table 7 are percentages of a "scientifically" selected nationwide sample of 629 respondents.

How accurate would you expect the opinion estimates to be in the 4 categories listed? Specifically, how accurate (inferentially) is the 60% figure that is reported for respondents with an eighth grade education? (Any background influences of importance here?)

8.  An article in a newsmagazine stated, "Allowing for 3% sampling error, the results can be projected to the total adult population of the U.S." (The sample size was 778.) How should this statement be interpreted?

9.  Six of the nearly 950 delegates to the 1972 Democratic convention were surveyed in 1975 concerning their opinions of the outlook for the 1976 convention. How accurate are percentage estimates likely to be? (Review possible background influences.)

10. During the mid-1970s mathematical modeling of the United States economy was very difficult. Never before had both recession and inflation simultaneously influenced the economy.

Economists found it very difficult to predict the direction in which the economy was heading. One newspaper said: "Although the council [the President's Council of Economic Advisors] customarily forecasts for the gross national product and several of its main components in quantitative terms, the report warned: 'The uncertainties are so great at the present time that the projections cited, although presented as specific numbers, are subject to an unusually wide margin of error.' "

What are the consequences of an "unusually wide margin of error" as far as interpretation of such statistics are concerned? (Such an estimate is sometimes called a "guesstimate.")

11.  Discuss the error of estimates for a Gallup, Harris, or other national opinion survey that appeared in a newspaper recently. How is your interpretation of the estimates affected by the background influences we have discussed?

12.  Gregor Mendel ran a number of tests of a hypothesis regarding whether the results of his many sweet-pea experiments followed prescribed ratios. For example, if two plants with a heterogeneous dominant characteristic, denoted **Aa**, are crossed, the offspring should be of the type **AA**, **Aa**, and **aa** in ratio of 1:2:1. The tests Mendel ran didn't involve a normal model; they involved a $x^2$ (chi-square) model, which we will discuss in chapter 9. Mendel's data was "too good," as might be illustrated by a normal analogue to his test results.

Let the diagram in figure 8 represent our critical area for a 5% normal test.

If this normal test were run independently, let's say, 1,000 times, how many times would we expect to reject the null hypothesis, even when it is true? How would you feel if all 1,000

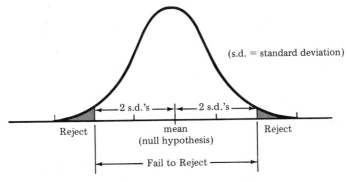

FIGURE 8    **Pictorial Representation of a 5% Test of Hypothesis**

tests led to failure to reject the null hypothesis? In essense, *all* of Mendel's tests of a (correct) null hypothesis led to failure to reject the null hypothesis.[7]

13. Are the error estimates for the sample proportions given in this chapter good for voluntary response surveys or low response surveys?

14. Discuss statistical significance at it relates to the FTC ruling mentioned in the section "Testing a Statistical Hypothesis" in this chapter.

15. A household that "views" public television is one in which at least one member of the household watched at least one show during the survey week. A respondent is usually asked to verify an affirmative response to a survey question on viewing by naming and describing the show that was viewed. This definition of viewing is different from the definition of viewing commercial television. Discuss this definition of viewing relative to commercial television. Should the viewing of public television be different from the viewing of commercial television? Is verification of yes responses necessary?[8]

16. Journals that publish research results have a tendency to publish only the reports of significant results. To illustrate the impact of such a publication policy consider an experiment that compares two drugs. Let us assume that 20 different researchers ran similar experiments comparing the two drugs. Suppose that one of the 20 researchers found a significant difference between the drugs. If all 20 researchers tried to publish the results of their experiments, it is likely, due to the above publication policy, that only the single researcher reporting a significant result would get his report published. Assuming that all 20 researchers used a 5% test, is this published result more likely to be the product of a Type-I error or the result of a real difference between the drugs? The above publication policy has what effect on the research reports professionals read in journals?

# Further Readings

There are a number of interesting articles concerning statistical inference in Tanur's *Statistics: A Guide to the Unknown,* cited in the chapter Notes. Articles by Link and Chapman are noted in this chapter. Now that you have some insight into the interpretation of inferential statistics, the articles in this fine collection should, for the most part, be quite easy to read and understand.

# Notes

1. R. F. Link, "Election Night on TV," *Statistics: A Guide to the Unknown,* ed., J. M. Tanur (San Francisco: Holden-Day, 1972), pp. 137–145.
2. Reprinted by permission from TIME, The Weekly Newsmagazine: copyright Time, Inc., 1971.
3. C. A. Bennett and N.L. Franklin, *Statistical Analysis in Chemistry and the Chemical Industry* (New York: John Wiley & Sons, Inc., 1954), pp. 246–247.
4. I. Guttman, et al., *Introductory Engineering Statistics* (New York: John Wiley & Sons, Inc., 1971), p. 267.
5. Main, "The Effect of Familiarity with the Examiner upon Stanford-Binet Test Performance," *Contribution to Education*, 381 (New York: Teacher's College, Columbia University, 1929).
6. D. G. Chapman, "The Plight of the Whales," *Statistics: A Guide to the Unknown,* ed., J. M. Tanur (San Francisco: Holden-Day, 1972), pp. 84–91.
7. J. L. Hodges, *Statlab* (New York: McGraw-Hill, 1975), pp. 228–229.
8. Jack Lyle, *The People Look at Public Television* (Washington, D.C.: Corporation for Public Broadcasting, 1975).

# 7 Epidemiology

STATISTICIANS WORK WITH RESEARCHERS in many different areas. Statistics furnish researchers with the tools necessary to properly design, execute, and analyze an experiment. No area of research is more likely to directly affect each of our lives than the work of the epidemiologist in medical research. It is for this reason that we take time at this point in our discussions to look at the role of the epidemiologist.

Epidemiology, the study of disease — its cause and prevention — is a very intriguing area of statistical research. Historically, the epidemiologist first studied infectious diseases such as smallpox, cholera, and yellow fever. Contemporary study involves the epidemiologist with noninfectious diseases such as cancer, heart disease, mental illness, diabetes, and accidents.

We will examine the observational and experimental research of the epidemiologists. In such research people may be observed through retrospective, prospective, or point-in-time studies. Actual experimentation on humans is very carefully regulated: experimentation on laboratory animals must first indicate that testing on humans is warranted.

A look at the causes of diseases such as cancer and heart disease involves us with one of the most difficult problems facing the modern epidemiologist. The problem is that the diseases studied today have many interrelated causes so that the effect of any one agent by itself is very difficult to determine. In this chapter we will examine this problem as well as some of the other aspects of epidemiology.

## What Is Epidemiology?

Epidemiology developed from the study of epidemic diseases (cholera, yellow fever, smallpox, for example).

> The word *epidemic* refers to the unusually frequent occurrence of a disease.

From the study of epidemics it became clear that the study of endemic disease was necessary.

> The word *endemic* refers to the usual frequency of occurrence of a disease.

Infectious diseases were of greatest concern to the epidemiologist prior to the twentieth century. Present emphasis is with the study of the noninfectious diseases such as cancer and heart disease.

> *Epidemiology* is the study of the cause(s) of a disease, how a disease is transmitted in a population, and, ultimately, how the disease can be prevented.

After a look at a history of epidemiology, we will look at the main approaches to epidemiological research — observation and experimentation.

## A History of Epidemiology

Since it is concerned with the study of disease, epidemiology must go back to the time when people first observed sickness and wondered why some people got sick and others did not. The formal study of the spread of disease in a population began with Hippocrates (about 460–377 BC). Regarded as the first true epidemiologist, Hippocrates was the first to attempt to characterize disease on a rational rather than a supernatural basis.[1]

The consideration of the contagion of diseases began with Hierouymus Fracastorius (1478–1553). In 1546 he described the transfer of infection by means of minute, invisible particles.[2] Although this represented the first formal expression of the concept of contagion, the treatment of leprophobes (lepers) indicated an informal knowledge of contagion. (Leprosy is a chronic, communicable skin disease caused by a microorganism.) Before the time of Christ, lepers were provided a bell and cup and prohibited from contact with nonlepers. This was one of the first instances when diseased individuals were isolated from the nondiseased portion of a community. Isolation of tuberculosis patients is a more recent example. (See Rosen for a more detailed discussion of the very early work in the area of contagious diseases.[3]) But it was not until the middle of the eighteenth century that the theory of contagion was generally accepted.

In America during this time contagion may have been exploited, giving rise to an early instance of biological warfare. American Indians were given blankets from smallpox patients during the French and Indian War. Epidemics consequently developed in the Indian tribes.[4]

Important for the study of disease in a human population are data referred to as vital statistics. John Graunt's work with the *Bills of Mortality* in London was the beginning of vital statistics.

> *Vital statistics* are data relating to birth, death (mortality), marriage, divorce, and the occurrence of disease (morbidity).

These first life-tables were not only a basis for the beginning of the life insurance business but also for the beginning of observational epidemiology.

Immunization from infectious diseases began in the eighteenth century, opening this area of research to the epidemiologist. In 1774 Benjamin Jesty, a farmer, infected his wife and two of his sons with cowpox material. (The cowpox virus is used as a smallpox immunate, that is, by exposing a person to the cowpox virus, a person becomes resistant to the smallpox virus.) Subsequent injection of smallpox material did not produce smallpox, that is, Jesty's family had become immune to smallpox.[5] Before this time a person become immune to a disease only after actually having contracted the disease.

Despite Jesty's work, Edward Jenner (1749–1823) is usually given credit for the first smallpox vaccination using the cowpox virus. Jenner, whose work was done independently of Jesty's, is probably noted over Jesty for two reasons: first, Jesty's wife nearly died from her injection so Jesty's work was not widely disseminated; and second, Jenner's work was quite successful in design, proof of immunity, and acceptance.

Observational epidemiology became highly developed by the work of John Snow (1813–1858). Snow carefully observed outbreaks of cholera. (Cholera is both an endemic and epidemic infectious disease that is most commonly disseminated through contaminated drinking water.) He described its cause and gave very detailed instructions on its prevention. Snow theorized that cholera was transmitted from person to person by the ingestion of living cells found in infected fecal material carried in contaminated water.

A pump without a handle, as shown in figure 1, is the symbol of John Snow's work to determine the cause of cholera. During the 1849

FIGURE 1
Wooden Pump of the 1850s
*Without* a Handle

London epidemic, Snow observed a heavy concentration of cholera deaths around the Broad Street pump in Golden Square. After Snow talked to the board of guardians for the area, the handle of the pump was removed. It was not, however, until his great experiment of 1853 that Snow was able to prove his contention that cholera was carried by contaminated water.

During an 1853 epidemic in London he found a much higher mortality from cholera among people who got their water from the Southward Vauxhall Company. This water came from the sewage-polluted basin of the Thames River. A second group of people, next-door neighbors to the first group, got their water from the Lambeth Company. The Lambeth Company's source of water was also the Thames but from a location farther upstream that was not contaminated.

Snow's work is important also because of his use of a control group as well as an experimental group.

> An *experimental group* is made up of people who get a special treatment.
>
> A *control group* is a group that is like the experimental group in all respects except the treatment given the experimental group.

In Snow's experiment the control group was made up of people whose main water source was not contaminated. "Treatment" in Snow's experiment was the drinking of contaminated water. Note that in Snow's experiment the control and treatment groups were neighbors and alike in all respects save their main source of drinking water. Hence, if a difference in death rates between these two groups was found, the treatment could be pointed to as the cause. Table 1 summarizes Snow's findings. Based on the information presented in table 1 Snow concluded that contaminated water caused cholera.

Modern epidemiological studies are less concerned with infectious diseases. Today the emphasis of epidemiological investigation is with diseases that are presumed to be noninfectious. Examples include coronary heart disease, diabetes, accidents, cancer of all types, and mental illness. Since these diseases can affect all of us directly or indirectly, the importance of having some knowledge of epidemiology is obvious.

**TABLE 1    Cholera Deaths, London 1853**

|  | Number of Houses | Deaths from Cholera | Deaths in each 10,000 Houses |
|---|---|---|---|
| Experimental Group (Water Source: Southward Vauxhall Co.) | 40,046 | 1,263 | 315 |
| Control Group (Water Source: Lambeth Co.) | 26,107 | 98 | 37 |
| Rest of London | 256,423 | 1,422 | 59 |

Source: J. Snow, *Snow on Cholera* (London: Oxford University Press, 1936).

# Observational Research

> In *observational studies* in epidemiology, investigators observe events in their natural setting, being careful to do nothing that might influence the occurrence of the events of interest.

We've talked before about the difficulties of trying to observe any population in its entirety. The sampling methods discussed in chapter 5 may therefore be used by epidemiologists.

Since interest in epidemiological studies is with diseases, a survey is likely to be made of doctors' offices, clinics, or hospitals in an area rather than of households or of people in an area. For large areas a sample of medical facilities might be contacted; for a smaller area all the medical facilities may be contacted (a census). We often hear about such surveys in the news media. During the flu season we hear about the number of reported cases of the flu for that year. The word "reported" indicates that information was obtained from medical personnel in the area.

There are times when people rather than medical facilities are contacted in epidemiological surveys. Sometimes researchers wish to obtain information on certain diseases early in their development — for example, diabetes and cancer of the cervix, which can be detected at early stages. In these situations a survey of people, rather than a talk to medical personnel in an area, is necessary if accurate information is to be obtained.

In this section we will discuss various types of surveys used in observational research. Whatever the type of survey, an epidemiologist is interested in observing either *morbidity* (disease) or *mortality* (death from a disease).

## *Point-in-Time Survey*

The surveys discussed in chapter 5 are called point-in-time surveys.

> Although it takes a certain period of time to run a survey, a *point-in-time survey* involves taking a sample from the population as it exists at a particular time.

For example, the surveys run in conjunction with the census represent data as of a set time, 1 April of the census year. Opinion surveys, preelection polls, television-viewing surveys, and marketing research surveys are other examples of point-in-time surveys.

In addition to the point-in-time survey in epidemiological research there are two other types of surveys that are to be discussed in this section.

### Prospective Study

Throughout our discussions of observational studies in epidemiology we will be concerned with two types of factors. The *antecedent factor* is an agent that precedes the onset of a disease. An epidemiologist is interested in studying various antecedent factors that might cause a disease. For example, in a study of heart disease an epidemiologist would be interested in studying many antecedent factors such as diet, exercise, smoking, family history, and life-style.

The other type of factor is the *outcome factor.* We will think of this factor as a dichotomy — "disease" or "no disease." This factor necessarily follows the antecedent factor, which will also be considered as a dichotomy — the antecedent factor is "present" or "absent." People may be selected for observational studies according to the presence or absence of either the antecedent or the outcome factors.

> In a *prospective study* people are selected according to the presence or absence of the antecedent factor and then observed over an extended period of time in order to monitor the development of disease.

In a prospective study we are observing two populations rather than a single population as is the case for a point-in-time survey. The two populations represent people with and people without the particular antecedent factor of interest. For example, in a study of the effects of oral contraceptive use on women the two populations of interest would be women who do and women who do not take oral contraceptives. These two populations would be observed over a period of time. The proportion of women in each group who develop the disease of interest (circulatory disease in this case) is noted. We would then ask if the proportion of women developing circulatory disease in the group of women who use oral contraceptives is the same as the percentage of women developing circulatory disease who do not take birth control pills. The fact that studies of this type have found that a greater proportion of women using oral contraceptives get circulatory disease than women who do not use oral contraceptives has caused many physicians to reevaluate whether or not to prescribe birth control pills for their female patients. Women who have been using oral contraceptives for a long period of time, especially women who smoke, may be advised to use other means of birth control.

There are advantages and disadvantages in taking a prospective look at people. An advantage is that we are able to estimate the risk or probability that a person will develop a particular disease. In the oral contraceptives example we could, with a prospective study, estimate the chances that a woman will develop circulatory disease if she does or does not take birth control pills. This is important information, and a comparison of women who take oral contraceptives with women who do not is possible. Such a comparison is an advantage in a determination of whether an antecedent factor causes a disease.

There are three major disadvantages of a prospective study. First, the study must be large and is therefore very expensive. We are looking at the proportion of people who get a disease. This proportion is usually very small. Therefore, many people must be studied so that a reasonably large number of diseased people can be observed. For example, about 4.5 of every 100,000 nonsmokers die from lung cancer each year. We would have to study over 2,000 nonsmokers for one year before we would expect to observe one fatal case of lung cancer. Since the rate of many diseases is small, prospective studies involve the observation of hundreds of thousands of people.

A second disadvantage of a prospective study is that a great deal of cooperation from participants is needed. The question arises whether those who volunteer to participate and actually do participate throughout the study are in any sense different from nonparticipants. This problem is not unlike the problem we discussed with voluntary response surveys. In the latter case we were concerned with whether respondents differed from nonrespondents.

A third disadvantage of the prospective study is that of self-selection. People choose which sample they want to be in. For example, in the oral contraceptives example women choose to use birth control pills. The question arises whether women who choose to use oral contraceptives are different from women who choose not to. Could there be a second antecedent factor that distinguishes users from nonusers? This second consideration might lead to both use of birth control pills and development of disease.

Another example that illustrates the self-selection problem involves smoking as an antecedent factor. Are smokers, let's say, more nervous than nonsmokers, and is it this nervousness that leads people to both smoke and to get disease? Self-selection is an important problem as differences between groups with and without one antecedent factor may be due to differences with regard to another antecedent factor.

Even with the three disadvantages we have discussed, the prospective study has become an important, but possibly overrated, tool in the proof of causation.

### Retrospective Study

As we have discussed, one of the disadvantages of a prospective study is its size. Due to the low rate of occurrence of a disease, many people have to be observed so that a few diseased people can be studied. To remedy this problem people who have the disease in question might be selected. An epidemiologist then looks into the backgrounds of these people to see what antecedent factors exist.

> In a *retrospective study* the investigator looks into the background of people who have a disease in order to determine potential causes of the disease.

In order that potential antecedent factors can be properly evaluated a control group is also selected. The control, or nondiseased, group is compared with the diseased group as any two populations might be compared. In a retrospective study the two populations studied are diseased people and nondiseased people, and for a prospective study the two populations studied are people with and people without a particular antecedent factor.

The main advantage of a retrospective study is that it is small in size. Since we start with people with the disease, we can study a relatively large number of diseased people with a relatively small overall study.

There are two main disadvantages of a retrospective study. First, it is difficult to point an accusative finger at a single antecedent factor. The diseased and nondiseased groups may differ according to many antecedent factors. Also, since an estimate of the risk of getting a disease cannot be made, a retrospective study does little to prove causality.

A second disadvantage involves a problem of statistical inference. The control, or nondiseased, group often consists of hospital or autopsy cases, since diseased people are likely to be in a hospital or have died from the disease. But can inference be made from hospital or autopsy patients to the general public? Certainly, people in hospitals are different from the general public so that inference to the public can be weak.

A look at an example involving autopsy cases illustrates this point. In cases in which an autopsy was performed the victims were cross-classified by presence and absence of tuberculosis and cancer. Cancer was found to be less frequent in autopsy cases with tuberculosis. Pearl inferred that cancer could be arrested if tuberculin (the protein of the tubercle bacillus) was injected into cancer patients.[6] He did not recognize that unless all deaths are equally likely to be autopsied, inference to living patients is not proper. Here a strong association existed in autopsy cases but not in the general human population.

Although we have discussed retrospective studies after prospective studies, retrospective studies usually precede prospective studies. The retrospective study signals potential antecedent factors that can then be studied in some detail by a prospective study.

For example, oral contraceptives were flagged as a potential health hazard when women who had had heart attacks were compared to controls. There existed a greater proportion of pill users among the heart attack victims than among the controls. Prospective studies then pinpointed the increase in risk due to pill use. (Women who use birth control pills develop circulatory disease at a rate of 28 per 100,000; women who don't take the pill get circulatory disease at a rate of 6 per 100,000.[7])

## Comparison

Let us consider once again the differences between point-in-time, prospective, and retrospective studies. Again we assume a dichotomy for both the antecedent and outcome factors: people do or do not possess the antecedent factor of interest and these same people do or do not have the disease in question. Subjects can then be classified to a two-way *contingency table,* as illustrated in table 2. People counted in box A in table 2 possess the antecedent factor and have the disease in question; people counted in box C have the antecedent factor but did not get the disease; and so on.

TABLE 2    Two-Way Contingency Table

|  |  | Antecedent Factor | |
|---|---|---|---|
|  |  | Yes | No |
| *Outcome* | Yes | A | B |
| *Factor* | No | C | D |

A point-in-time study is characterized by people being selected from one population. Each subject is then counted in one of the 4 boxes of the contingency table.

For a retrospective study people from a diseased population are counted in box A or box B of table 2. Subjects in the control, or nondiseased, group are counted in box C or box D. Note that the number of people with the disease (the number of people in boxes A and B) is fixed ahead of time as is the number of controls (people in boxes C and D). This contrasts to the point-in-time study in which the total number of people in all 4 boxes is the only number that is preset.

In a prospective study people with the antecedent factor are placed in box A or box C and people without the antecedent factor are placed in box B or box D. Here the total number of people in boxes A and C is set before hand as is the total number of people in boxes B and D.

So we see that these three types of surveys fix various sample sizes to be selected from different populations.

### Examples

It might be instructive at this point to give the details of both a retrospective and prospective study. We will look at two studies on smoking and lung cancer, which influenced the famous 1964 United States Surgeon General's report on *Smoking and Health.*

Doll and Hill ran a retrospective study on lung cancer.[8] In this study 1,400 lung cancer patients were age- (within 5 years) and sex-matched with a control group of patients with a respiratory disease other than lung cancer. The lung cancer group was found to consist of heavier smokers than the control group. (Doll and Hill found similar results using community controls.)

Note the age and sex matching of the diseased and control groups mentioned. An investigator would not want the diseased and control groups to vary in age distribution as age could be a potential antecedent factor; older people are more likely to have a particular disease. The same reasoning makes sex matching desirable. It is desirable to match for as many potential antecedents as possible so that the diseased and control groups differ by one antecedent factor at most.

A retrospective study on the smoking and health question cannot, however, give any indication of the extent to which smoking might increase the likelihood of a disease such as lung cancer. A prospective study is required for such an estimate of risk. In a prospective study comparable groups of healthy smokers and nonsmokers are selected and watched over time. (Recall that in a prospective study the factor thought to be related to the disease is studied, and in a retrospective study people who have the disease are selected.) The Dorn study of

United States government life-insurance policyholders is a prospective study that was conducted on the question of the relationship of smoking and health.[9] Dorn obtained 85% (250,000) response on two mailings of a smoking questionnaire. By obtaining death certificates from the Veterans Administration Dorn and his associates were able to determine the cause of death for each person who died during the 8.5 years of study of these respondents. For people who died of lung cancer, Dorn found a *mortality ratio* of 11 for cigarette smokers versus nonsmokers. This figure is the rate of death in a year from cancer of the lung among smokers divided by the rate of death in a year from lung cancer among nonsmokers. Hence, in a given year a person who smokes is nearly 11 times more likely to die of lung cancer than a nonsmoker.

TABLE 3    Mortality Rates per 100,000 each Year

|  | Smokers | Nonsmokers | Mortality Ratio |
|---|---|---|---|
| Lung Cancer | 48.3 | 4.5 | 10.7 |
| Coronary Artery Disease | 294.7 | 169.5 | 1.7 |

Source: J. Fleiss, *Statistical Methods for Rates and Proportions* (New York: Wiley, 1973), p. 60.

## Interpretation

A word of caution when interpreting mortality ratios is necessary. Rates, like percentages, must be interpreted with regard to the number of people involved. For example, the mortality rate for lung cancer among smokers cited in *Smoking and Health* was about 48, that is, 48 in 100,000 died of lung cancer in one year.[10] Hence, less than 1 in 2,000 smokers are likely to die of lung cancer in a year. This figure is not as enormous as the factor of 11 just discussed.

We must be careful when comparing different mortality ratios. The data in table 3 are from prospective studies reviewed in the 1964 *Smoking and Health* report of the surgeon general. (These data are a composite of a number of studies.) From table 3 we see that a smoker is about twice (actually 1.7 times) as likely to die of coronary artery disease than a nonsmoker. Compare this to the factor of almost 11 for lung cancer. A tenfold increase over a rate of one per hundred thousand is equivalent to a tenfold increase over a rate of one per hundred!

$$\text{mortality ratio} = \frac{\text{mortality rate in group with antecedent factor}}{\text{mortality rate in group without antecedent factor}}$$

$$\frac{48.3/100,000}{4.5/100,000} = 10.7 = \frac{48.3/100}{4.5/100}$$

The mortality ratio does not take into account how many people died from a disease, just the ratio between groups. Many, therefore, maintain that the actual difference between rates is more meaningful since it reflects how many more (not how many times more) people are likely to die from a disease. Calculating the difference between rates from table 3, smoking would be responsible for an increase of about 295 − 170, or 125 deaths per year from coronary artery disease per 100,000 people. On the other hand, smoking would be responsible for an increase of about 48 − 4, or only 44 deaths per year per 100,000 people — that is, heart disease should be the more serious concern of smokers.

One advantage of a prospective study is that the difference between the ratio of diseased people with and without an antecedent factor can be estimated as we estimate the difference between proportions (see chapter 6). This difference cannot be estimated in a retrospective study. However, both studies give an estimate of the mortality ratio. If we would like to compare the results between a retrospective study and prospective study, the mortality ratio would be used.

$$\begin{array}{c}\text{mortality ratio} \\ \text{(prospective estimate)}\end{array} = \dfrac{\begin{array}{c}\text{mortality rate in} \\ \text{antecedent group}\end{array}}{\begin{array}{c}\text{mortality rate in group} \\ \text{without antecedent factor}\end{array}}$$

$$\begin{array}{c}\text{mortality ratio} \\ \text{(retrospective estimate)}\end{array} = \dfrac{\begin{array}{c}\text{rate of antecedent factor in group} \\ \text{who died from disease}\end{array}}{\begin{array}{c}\text{rate of antecedent factor in group} \\ \text{who died without the disease}\end{array}}$$

$$\times \dfrac{\begin{array}{c}\text{rate without antecedent factor} \\ \text{in group who died without disease}\end{array}}{\begin{array}{c}\text{rate without antecedent factor} \\ \text{in group who died from disease}\end{array}}$$

The two equations relative to mortality ratio are valid if the disease rate is low and if the nondiseased and diseased groups in the retrospective study are random samples from corresponding populations. (A corresponding morbidity ratio can be defined for rate of occurrence of a disease rather than death from a disease.) Because retrospective and prospective studies can be compared using the mortality ratio, this statistic is used more often than the difference between the mortality rates in groups with and without the antecedent factor, values found only from prospective study. Be aware, however, of how informative the difference can be, and remember, the ratio can be misleading.

# Experimentation

Experimental research concerns the study of a population that is created solely for the purpose of the investigation. An epidemiologist can approach a research problem through controlled experimentation as well as through the kind of observation just discussed. Our discussion here will be directed toward the problems an epidemiologist encounters when conducting human and animal experiments. The design of experiments will be discussed in the next chapter.

## *Ethical Questions in Human Experimentation*

Experimentation on humans is very carefully controlled. The ethical questions involved with experimenting on humans are so profound and complex that at present experimentation with humans must be approved so that the rights of the people used in the experiment are not abused. Ethical considerations of human experiments raise the questions:

1. Should an experiment deliberately expose someone to an agent suspected of causing a disease?

2. If a treatment is thought to be effective in treating a disease, should some people *not* be given the treatment?

3. Should ill people be experimented on, or should every attempt be made to heal them?

Let us look at three examples of human experimentation that will bring these ethical questions into sharper focus.

First, consider a test of the Finlay hypothesis. Carolos Finlay of Cuba hypothesized that yellow fever was carried by a particular type of mosquito (***Aedes aegypti***). The Walter Reed Commission set out to test this theory between 1900 and 1901.[11] Human volunteers were subjected to various types of possible exposure to yellow fever. The Finley hypothesis was confirmed, since those exposed to the bite of the **A. aegypti** developed yellow fever. (The mosquito obtains the microorganism that causes yellow fever by first biting a person who has yellow fever.)

Should humans have been exposed to this suspected yellow-fever agent?

Second, consider the testing of the Salk vaccine as an immunate against polio, a disease causing paralysis that was somewhat common in the 1950s. Two types of viral vaccines are used — live and "dead" virus vaccines. Immunization by means of live virus vaccines such as the cowpox vaccine for smallpox involves the injection of a live virus into a person. This virus will not produce the undesirable disease (smallpox in the case of cowpox vaccination) but will produce the antibodies that fight the disease in question. The reaction is usually a minor one. A dead virus vaccine is one in which the strength of the virus is reduced, usually with formaldehyde, to a level that is not supposed to produce the disease but will produce desired antibodies to fight off future viral infection. The Salk vaccine being discussed is a dead virus vaccine. Current polio vaccines are live virus vaccines.

A part of a very large study involved about 750,000 first, second, and third graders.[12] Each child was vaccinated. Some of the children were injected with a salt water solution, and others received the Salk vaccine. A random mechanism was used to decide which child got which injection. Since no bottle was marked, no one knew who got which injection.

Should all children have been given the vaccine if it was thought to be an immunate against polio?

The third human experiment we will consider was conducted by James Lind to find a cure for scurvy.[13] (A lack of Vitamin C causes scurvy, which was especially prevalent among sailors.) Twelve patients with scurvy were put to sea in a ship (1747). All 12 received three good meals a day plus one other treatment: two got cider; two got *elix vitrial* (aromatic sulfuric acid); two got vinegar; two got seawater; two got citrus fruit (two oranges and one lemon); and the last two got an electuary (a medical preparation) recommended by a hospital surgeon. The two receiving citrus recovered and tended the others.

Should experimentation be conducted on the ill?

### Animal Experimentation

Experimentation on animals does not usually raise the consciousness of people as does human experimentation, although such experimentation is also monitored. An interesting problem of inference arises with such experiments. Can one infer to the human population from experiments conducted on a sample of, let's say, rats? A substance is considered carcinogenic (cancer-producing) if it produces cancer in laboratory animals. What can be said about the effect of a substance on humans when it is known to be cancer-producing in rats?

The problem of inference is further compounded by the massive doses of a drug that animals are given in laboratory tests. With small doses, doses equivalent by body weights to levels consumed by humans, two problems arise. First, a large number of mice must be tested since only a few will develop cancer. Second, the experiment will be of long duration (years) since it takes time for cancers to develop.

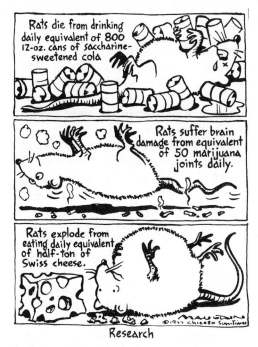

Research

In order to reduce the number of animals to be tested and the
duration of such experiments animals are given massive doses. (Doses
are massive in comparison to the amount humans would normally
consume.) The disadvantage of these large-dose experiments is that a
laboratory animal may develop cancer because of the massive dose. An
animal's body may be able to pass or metabolize small amounts of a
compound, but the animal's natural defense mechanisms may be overwhelmed by a large dose; that is, a dose equivalent to the levels of a
drug usually consumed by humans may have no effect although massive doses may cause cancer.

Compound this problem with the fact that the bodies of laboratory animals react differently to a drug than the bodies of humans.
Even with these questions of inference, many feel that a drug that
produces cancer in laboratory animals at *any* dosage should not be
consumed by humans.

A combination of animal and human studies are needed. For
example, saccharin fed to rats at massive levels produced bladder tumors. A retrospective study in Canada found that men who had bladder cancer had consumed more artificial sweeteners.[14] Such a retrospective study should be followed by a prospective study. Then, an
overall picture of the effects of saccharin on humans can be painted.

## Causation

Proving that a particular agent causes a disease is seldom an easy undertaking as we have just seen. Difficulty arises from the possibility that several agents interact to cause a disease. Heart disease is a case in point. Many factors are known to affect the condition of the human heart — amount of daily exercise, blood cholesterol, weight, and smoking habits, to mention a few.

Causation is also hard to establish since human experiments are usually required. To prove, for example, that smoking causes lung cancer we would select a very large sample of people and randomly assign them to one of two groups. One group would be the control group and would not smoke, and the other group would smoke. After a few years, we determine whether a larger proportion of smokers than nonsmokers develops lung cancer. Clearly, such an experiment is ethically impossible as well as impractical. How practical would it be to try and supervise such an experiment?

In some instances a third factor related to the suspected causative agent is responsible for both the agent and the disease. For example, endemic goiters increase in frequency with altitude. (A goiter is an enlargement of the thyroid gland causing a swelling in the front of the neck.) But altitude does not cause endemic goiters. These goiters result from low iodine content in food, a condition that is more frequent at higher altitudes.

Also possible is a third factor relative to a person's constitution or physical makeup. The hypothesis concerning a person's constitution, or the constitutional hypothesis, in the smoking and health area is the statement that a third factor (genetic or psychological, for example) makes a person more likely to smoke and more susceptible to, let's say, lung cancer. Under the constitutional hypothesis a person's constitution is such that he or she is more likely to smoke and more likely to get lung cancer. The genetic third factor has not been totally disproved although its exact interrelationship with smoking is not known. The surgeon general's report of 1964 noted that both twins of identical twins (people with the same genetic makeup) smoked more often than both twins in nonidentical pairs. Such data support the genetic third-factor theory. The rise in per capita smoking rates over the first half of the twentieth century discounts the genetic hypothesis, however. It is unlikely that genetic changes would simultaneously occur in all the countries that showed the increase in per capita cigarette consumption. Also, it is not likely that the changes would occur in such a short period of time.

The causal relationship between smoking and lung cancer was determined by the surgeon general's advisory committee of 1964 by the following:

1. Consistency of the association over so many diverse studies

2. Strength of the association (e.g., the 10.8 mortality ratio reported by Dorn)

3. Specificity of the association (the precision at which one variable, smoking, predicts another, cancer of the lung [see chapter 9]

4. Temporal relationship of the association (Smoking should precede, temporarily, the onset of the disease.)

5. Coherence of the association with known facts in the natural history and biology of the disease (One aspect of the coherence of the association between smoking and lung cancer is that many known carcinogens are found in tobacco smoke.)

The controversy continues, however. It is important to note here than an association between two measurements or variables does not necessarily establish a cause-and-effect relationship between the variables. An example of such an association is the increase in the incidence of lung cancer that accompanies an increase in smoking. We will return to this in chapter 9.

Our idea that causation means a one-to-one relationship between a causative agent and the disease has to change. There is clearly not a one-to-one relationship between smoking and lung cancer. Some people who smoke never get lung cancer, and people who never smoke do get lung cancer. Causation may have to be redefined. It may be possible that a person has, let's say, three factors required for the formation of lung cancer. Any three of many different factors are necessary for the development of lung cancer. Any of the many possible factors can be said to be causative because if it were eliminated, the disease is less likely to occur. But of course the disease could still occur if three other factors are present.

Genetics cannot be ruled out of the cancer picture. One theory, for example, says that a person's genetic makeup would determine if 3, 4, or even 5 factors are needed for a cancer to develop. We can expect to see much research being done on the relationship of genetics and cancer development.

# Summary

We have taken a very brief look at the intriguing work of the epidemiologist. An epidemiologist studies infectious diseases (polio, smallpox, tuberculosis, etc.) as well as noninfectious diseases (cancer, accidents, mental illness, heart disease, etc.). Epidemiologists seek to determine the cause of a disease, how it is transmitted in a population, and, ultimately, how the disease can be prevented. Epidemiological research is observational or experimental in nature.

The main types of observational studies are retrospective and prospective studies. In a retrospective study individuals with the disease are studied. Factors that are different between diseased individuals and nondiseased controls can then be determined and studied further. In a prospective study people are observed over an extended period of time. The strength of the association between a disease and possible causative agents can then be determined.

Experimentation in epidemiological research involves experimentation on humans and laboratory animals. Human experimentation was discussed in some detail. Many ethical questions about such experimentation were raised. Animal research raises questions of inference. Can one infer to the human population from an experiment conducted on a sample of animals?

The proof of causation is probably one of the more difficult problems facing the epidemiologist. In the cases of cancer, heart disease, accidents, and mental illness there are many interrelated causative agents. For example, smoking and genetic factors may both be involved in the occurrence of lung cancer. There may be some people who are genetically more susceptible to cancer and who should not smoke, and there may be others who are less susceptible to lung cancer who may not increase their chances of lung cancer by smoking.

Having introduced you in this chapter to the observation of an existing population as well as to the study of created populations in experimentation, in chapter 8 we will look at how to design a study of a population. An appreciation of how to design an investigation of a population, existing or created, should aid our understanding of the statistical analyses resulting from such studies.

# No Comment

**"A GOOD THOUGHT . . .**

■    *"It is now proved beyond doubt that smoking is one of the leading causes of statistics."*
            *(Fletcher-Knebel,* Lexington Leader, *21 January 1976)*

■    *"Statistics released by the National Institutes of Health indicate that the leading cause of disease among laboratory animals is research."*
            (Parade Magazine, *5 February 1978)*

# Exercises

1. If someone wanted to run an experiment to prove that smoking caused lung cancer, how large a sample would be necessary? Would such an experiment be practical? (Consider both the size of the experiment as well as the problem of supervising such an experiment.)

2. Vinyl chloride was taken off the market in the mid 1970s when 26 cases of a rare liver cancer (angiosarcoma) were almost all traced to factories producing vinyl plastic. Some victims were workers in these plants, others lived near such plants. Is this a retrospective or prospective look at a rare disease (only 25 in 1.5 million die from it)?

3. The incidence of cancer deaths in the United States increased 5.2% from 1974 to 1975. (Usually this rise is about 1%.) What might cause this increase? What would an epidemiologist look for? Would a map indicating the change in rates for different areas of the country be useful?

4. We discussed poststratification in chapter 5 and chapter 6. Let's take a look at it again with respect to age-adjusted mortality rates. Poststratification by age is necessary since we cannot usually isolate people into different age groups and then randomly select within each group. In our case the mortality rates for both the treated group, for example, the smokers, and the control group, the nonsmokers, might be adjusted to the proportion of people in the entire adult population in each age category.

   Another procedure involves adjusting the treatment group to the control group. For example, suppose people whose age is between 30 and 39 make up 18.7% of the control group. The mortality rate for this age group in the smokers, or treated, group would be weighted by .187. Age-specific mortality or morbidity rates or ratios result.

   Is this adjustment necessary? Are the control and treated groups in a study on smoking and health the same except for treatment (smoking)?

5. One of the arguments contradicting the constitutional hypothesis in the smoking and health controversy is the sharp increase in (lung) cancer deaths in the first half of the twentieth century. This increase was paralleled by a simultaneous increase in

cigarette consumption. Consider possible background influences on cancer death rates from 1900 to 1950. Focus attention on these considerations:

a.  With cures for many infectious diseases being developed during this time as well as an increase in the quality of health care, people lived longer. Maybe if people in the early twentieth century lived longer, they too might have developed cancer. (Most cancers take 20 to 30 years to develop.)

b.  Death certificates are a major source of data on death from cancer. How accurate is this information? Note that when a disease is thought to be prevalent in an area, a doctor may be more likely to list that disease as the cause of death. Would an increase in cancer deaths then appear greater than it really was?

6.  How would an epidemiologist determine if the occurrence of a disease had reached epidemic proportions? What type of statistical inference would be employed?

7.  Consider the mortality data given in table 3 of this chapter. What is the probability that a smoker will *not* die of lung cancer in a year? Heart disease?

# For Discussion

1.  Because of the time required for a cancer to develop, 20–30 years in humans, the level of a chemical given to a laboratory animal is not equivalent to the corresponding level to which a human can be expected to be exposed. (A "corresponding level" would be a set concentration per unit of body weight.) Levels of a chemical will be increased so that the time to tumor development can be shortened. For example, a chemical used to decaffeinate coffee beans was found to be carcinogenic. For a human to get an equivalent daily dosage to what the mice in the experiment received he or she would have to drink 50 million cups of coffee a day over his or her whole life! What problems of inference arise here? Would an animal react differently to high levels of a chemical over a short time period than to low dose levels over a long period of time? Discuss this important problem.

2.  Discuss the measurement (count) of a disease by examining reported cases. Recall our discussions of reported crime in chapter 2.

3. A survey concerning the common cold would necessarily be targeted to households rather than medical personnel. Why?

4. We've talked of using volunteers in prospective studies. Are volunteers representative of a larger collection of people for, let's say, a study of the smoking and health question?

5. Discuss what safeguards would ᵇe needed to ensure that the rights of humans used in experiments are protected. Do problems of inference exist because participants must be volunteers?

6. Discuss the measurement of the consumption of smoke from cigarettes, pipes, and/or cigars. (Recall our brief look at ths problem in chapter 2.)

   In many prospective studies people have necessarily smoked cigarettes for varying periods of time. A person who smoked two packages of cigarettes a day for 10 years will often be equated to a person who smoked one pack for 20 years — that is, consumption is defined by *pack years*.

   $$\text{pack years} = \frac{\text{(packages of cigarettes smoked per day)} \times}{\text{(years smoked)}}$$

   Many different conversions of cigar and pipe smoking to cigarette smoking have been used. For example, one study equated one cigar to five cigarettes and one pipeful of tobacco to 2.5 cigarettes.

7. In 1976 President Ford suggested that all United States citizens be innoculated against a strain of swine flu, which was thought to be the flu strain for the upcoming flu season. This flu strain was thought to be highly contagious since all 6 people injected with this virus in an experiment contracted the disease. Five hundred cases (one death) had been reported at Fort Dix, New Jersey in the winter of 1975–1976. A form of swine flu swept the world following World War I, leading to 558,000 United States deaths — usually from pneumonia. (Such a worldwide epidemic is called a *pandemic*.) Discuss Ford's decision. (Recall our discussions concerning the making of decisions in chapter 4.) Observe that 1976 was an election year.

8. The Delaney amendment to the 1958 Food, Drug, and Cosmetics Act requires that any substance that produces cancer in laboratory animals or humans be banned. As we've discussed, saccharin was banned on a limited basis when a Canadian experiment

showed an increase of bladder tumors in rats fed saccharin. Five percent of the rats' diet was saccharin. This is equivalent to a human drinking 800 soft drinks sweetened with saccharin every day for his or her whole life. This led to a very careful look at this amendment. Should the Delaney amendment be changed? If so, how?

9.    Epidemiologists are often confronted with the problem of trying to trace the source of an outbreak of disease. Discuss how an epidemiologist might go about this task if, let's say, a number of people in a town develop food poisoning.

    In the case of Legionnaire's disease a number of people who had attended a convention in Philadelphia developed pneumonia (late summer 1976).[15] How would an epidemiologist go about finding the cause of this outbreak? Observe that, unlike the food poisoning example cited, the cause and mode of transmission of Legionnaires' disease were not known.

    Two weeks after the outbreak in Philadelphia no secondary infection had occurred — that is, the families of the legionnaires did not contract the disease from those who had attended the convention. How would this information affect the actions of the epidemiologist in the search for the cause?

# Further Readings

A look at epidemiology that does not get too technical can be found in the book *Epidemiology: Man and Disease* by J. P. Fox. It is listed in the Notes section of this chapter.

If you desire, the smoking and health controversy can be looked at in more detail. The surgeon general's report, *Smoking and Health,* Report of the Advisory Committee to the Surgeon General of the Public Health Service (Princeton: Van Nostrand, 1964) gives one side of the question, and *You May Smoke!* by C. H. Kitchen (New York: Award Books, 1966) gives the other side.

# Notes

1. J. P. Fox, *Epidemiology: Man and Disease* (New York: Macmillan, 1970), p. 20.
2. *Ibid.,* p. 21.
3. G. Rosen, *A History of Public Health* (Garden City, N.Y.: MD Pudlications, 1958).
4. L. D. Fothergill, "Biological Warfare and its Defense," *Public Health Report* 27 (1957): 865.
5. L. Clendening, *The Romance of Medicine* (Garden City, N.Y.: Garden City Publishing Co., 1933), p. 218–219.

6.  J. L. Fleiss, *Statistical Methods for Rates and Proportions* (New York: Wiley, 1973), pp. 7–8.
7.  Royal College of General Practitioners, *The Lancet* 2 (8 October 1977): 727. 727.
8.  R. Doll and A. B. Hill, "A Study of the Etiology of Carcinoma of the Lung," *British Medical Jurnal* 2 (1952): 1271–86.
9.  H. A. Kahn, *The Dorn Study of Smoking and Mortality among U.S. Veterans: Report on Eight and One-half Years of Observation.* (Washington, D.C.: National Cancer Institute, Monograph 19, 1966).
10.  Fleiss, *Statistical Methods for Rates and Proportions,* p. 59–61.
11.  W. Reed, "The Propogation of Yellow Fever. Observations Based on Recent Researches," Nedical Record 60 (1901): 201–209.
12.  P. Meier, "The Biggest Public Health Experiment Ever: The 1954 Field Trail of the Salk Poliomyelitis Vaccine." *Statistics: A Guide to the Unknown,* ed., J. M. Tanur (San Francisco: Holden-Day, 1972), pp. 2–13.
13.  J. Lind, *A Treatise on the Scurvy* (1753) (Edinburgh: Edinburgh University Press, reprinted 1953).
14.  G. R. Howe et al., "Artificial Sweeteners and Human Bladder Cancer," *The Lancet* 2 (17 September 1977): 578–581.
15.  D. W. Fraser et al., "Legionnaire's Disease," *New England Journal of Medicine,* vol. 297, no. 22 (1977): 1189–1197.

# 8 Design

KNOWING HOW TO DESIGN a sample survey can be an aid in interpreting survey results. This follows since knowing when a survey is designed so that credible information is found most efficiently will aid our interpretation of the survey results.

In this chapter we will look at sample survey design and also at the design of an experiment. We will see how an experiment might be run so that the results will be credible. As with our discussion of the design of a sample survey, our objective in discussing experiment design is to be able to interpret experiment results.

The topics of sample survey design and experiment design are discussed together because of the close relationship between experimentation and sample surveys. Sample surveys are involved with the investigation of an existing population. An experiment is concerned with a population created solely so that it can be studied. Many examples of experimentation and sample surveys exist in almost every area of scientific investigation. We will look at considerations unique to both sample survey design and experiment design as well as the considerations that are common to them.

## Sample Survey Design

> A *sample survey* is the study of an existing population. A 100% sample is called a census.

We have talked about judging the results of sample surveys throughout this book. Knowing how to judge the results of surveys should certainly aid us in considering the design of a survey whose results will be credible (and vice versa). It is desirable both as a review of interpretation of survey results and as a means of better judging surveys to see how we might design a good sample survey.

The purpose of these discussions is not to try to make you a pollster but to give you an appreciation of survey design. This appreciation is necessary for two reasons. First, it is likely that you may have become quite cynical about surveys and feel that no survey is worth your full consideration. This is not true. Although far too many surveys are poorly designed, you should be able to detect these and ignore the results. But many surveys such as the surveys run by the federal government are well designed and executed and contain much credible information.

Second, survey design must be discussed so that you will appreciate how difficult it is to run a good survey. Many people feel that a survey may be conducted by calling a few people on the telephone, by stopping a few people on the street, or by sending out some questionnaires. We will see that this view is completely wrong. Although a number will always result from any survey, our major concern is whether a credible statistic will result.

## Undesirable Types of Sample Surveys

Let us look first at types of sample surveys that do not necessarily lead to credible statistics.

### Convenience Sample

> A *convenience sample* is a sample selected in a manner that is convenient, but no attention is given to whether the sample is representative of the population of interest.

It would, for example, be convenient for a bus company to pass out a questionnaire to each of its riders. But the response rate of such a survey can be so low that inference to all riders is questionable and inference to all potential riders is out of the question. If the bus company is interested in people who might ride the bus but are not presently riding, a different sampling design is needed. Another survey design is called for if accurate information on present riders is desired.

Many of the sample surveys we have discussed can be classified as convenience samples. A politician who sends questionnaires to certain key constituents, a telephone company sending questionnaires with monthly bills, and the student government at a university giving questionnaires to members of certain dorms are all examples of convenience samples. We should not have to dwell on the credibility of the numbers resulting from these surveys.

### Judgment Sample

> A *judgment sample* is one that is determined by an "expert" to be representative of the population of interest.

For example, politicians may feel that to sample opinion on a subject they need to talk to at least one member of each of a number of representative groups, let's say, a conservative, a liberal, a black, and a Catholic. The politicians may sincerely feel that they are giving each group an opportunity to be heard. Variance of opinion within these groups as well as the size of groups relative to the subject can make the results of the judgment sample of questionable value. The size of the groups is likely to change as the topic of study changes; that is, a person might be a fiscal conservative, yet a liberal on matters of, perhaps, civil rights.

### Quota Sample

> A *quota sample* is one in which certain strata are proportionally represented in the sample.

Public opinion polls that are run today are, to some extent, quota samplings of the populace. Prior to 1948 opinion polls were based entirely on quota samples. Recall from our discussions in chapter 5 that such a sample can be biased toward the more educated segment of our population. Also, interviewers who must fill quotas are more likely to visit homes that are well kept. And, people agreeing to participate tend to be more educated than those who refuse to participate. Interviewing biases will affect resulting statistics in an unknown way, making interpretation of survey results difficult. Another problem with quota samples is that the uncertainty of statistical inferences is hard, if not impossible, to measure since a random mechanism is not used in the selection process.

Recall from our discussions of poststratification that weighting a poorly drawn sample by known strata sizes may give a sample that is at best a quota sample. Being sure that the number of people in a strata have their opinions reflected in the overall sample may be of limited value. The people talked to within each stratum must reflect opinion in that stratum. Poststratification weighting can give little aid to a stratum sample that does not properly represent the stratum.

Sample surveys that give credible results must be more than convenience, judgment, or quota samples. Let us see how a credible survey might be designed.

### *Designing Sample Surveys for Credible Results*

When designing a survey we must keep an eye on the likely results of the survey. We've talked about interpretation of survey results. The background influences on our interpretation must be considered before or at the time the survey is designed.

Recall from chapter 2 that we interpret survey results by considering:

1. Population sampled

2. Method of contact

3. Questions to be asked

4. Timing

5. Response rate

6. Sample size

7. Sample design

Let us consider each of these influences and see how we might ensure credible results with a little foresight.

**Target Population**  The population to be studied must be specified. This population is usually referred to as the target population. Deciding on the target population may also determine the method of contact for a survey. Let us look at some examples to see how this works.

Suppose we are contemplating a survey of voter preference prior to an election. The people whose opinions count are those registered voters who will vote. Ideally, we would like to select our sample from a list of registered voters. But, if the survey is intended to reach potential voters outside, let's say, a city, a list of all registered voters may be very difficult to obtain. In many cases these lists are not available until just a few days before an election, making them of limited use even if attention is restricted to a small geographic area. To overcome this problem, at least in the case of a preelection survey, unregistered as well as registered adults might be contacted. We might, then, use the telephone directory as a frame for a preelection survey. Thus, the method of contact has also been determined.

Another possibility would be for us to select from all households in an area rather than just the households with listed telephones. Since a list of all households in an area may not exist, we may decide on a multistage design. For a survey of a city, we might select blocks from a list of blocks in the city. In certain SMSAs block maps and summary statistics are available from census data. Otherwise, a list of blocks may have to be constructed. A personal interview survey is implied with such a target population. An interviewer must go to the selected blocks and choose households to be contacted, the second stage of the selection process. A systematic sample of households within a block is suggested, as an interviewer can easily be instructed on how to select such a sample and a supervisor can check to see if appropriate households were contacted.

In a survey of a larger area, perhaps a state or even an entire country, other stages would precede the selection of blocks. Towns of varying sizes and rural areas would first be selected with selection of blocks in towns and equivalent areas in the rural regions following. The next stage of selection would involve selection of households within blocks or designated rural areas. Stratification by rural area and size of urban areas is implied as selection within each of these strata would be different.

We quickly get the feeling, and rightfully so, that the determination of the target population may necessarily require the specification of many other design determinations. It does not take much reflection to see that the design of a survey, as just discussed, is indeed quite involved, especially when stratification by location of towns might further complicate the selection procedure.

**Method of Contact**    Another of the necessary design determinations is to decide on the method of contact to be used for the survey. Is the survey to be a mailback questionnaire, a personal interview, or a telephone survey?

As indicated previously, a determination of the target population might necessitate a particular method of contact. But we may have to decide the method of contact to use even after the target population has been specified. Cost and time constraints will often influence our decision about the method of contact — telephone, mail, or personal interview. Let us look at each method.

A telephone survey can be relatively inexpensive and can also be conducted quickly. Recall that the last poll taken before an election is likely to be a telephone poll that can be conducted in one day. Also, a telephone survey is relatively easy for the supervisor to control.

A mailback survey, on the other hand, will take more time to run than a telephone survey. We must give people to chance to respond. Remailings are necessarily involved as well as the possibility of having to contact nonrespondents or at least a subsample of nonrespondents if the response rate is low.

A personal interview survey is an expensive and time-consuming venture. Interviewers must be trained. Travel is expensive, since an interviewer is paid for time and mileage. On the positive side, more questions can be asked when a respondent is being questioned personally. However, this may be a disadvantage, as the tendency is to ask too many questions, making the accuracy of response low. I've seen one hundred-page questionnaires that take up to three hours to complete.

Whichever method of contact is chosen, we must keep in mind the interpretation of the results of each type of survey. A telephone survey may be less expensive than a personal interview survey, but what do we do about people without a telephone? Is information on that segment of the population important? A mailback survey may be relatively inexpensive and easy to administer, but what about nonresponse? These factors must be considered.

This is not to say that a telephone survey is bad because people without (listed) telephones are not contacted. On the contrary, many pollsters prefer telephone surveys since interviewing can be monitored. For personal interview surveys a pollster is not even sure an interview took place, let alone whether the interviewer did a good job. Moreover, good information is usually available from telephone surveys. Our concern here is that you be aware of the target population of such surveys. Recall that inference can be made only to the target population: inference to a larger population is, at best, tenuous.

**Response Rate**    In every survey design we must consider what will be needed to get the highest possible response rate. A premailing (letter or postcard) could be sent to the people in the sample so they will know that they will be contacted and questioned. People are more likely to respond if assured that no one is trying to sell them something. People also like to be assured of the confidentiality of their responses, which is another reason for a premailing.

For most surveys we must be prepared to recall people, that is more than one attempt to contact individuals who cannot be reached at first must be made. For a personal interview survey this means traveling to an area more than once. However, recall should not be at the same time of day for each visit or call.

Why bother with recalls? Suppose a telephone survey is conducted in the evening. People who work at night or are likely to be out in the evening may have opinions quite different from people who are likely to be home in the evening. Consider, for example, the age group from 18 to 25: people in this group are likely to be out in the evening; yet surely we would want their opinions in a preelection survey.

Whatever type of survey is being conducted, every effort must be made to ensure the highest possible response rate. Premailing and recalls are two ways to increase response. Another way to increase response is to train the interviewers well since caustic or poorly trained interviewers can only lead to lower response rates. This problem is avoided in mailback and telephone surveys, of course. With the mailback survey no interviewers are needed. And in the telephone survey we can monitor the interviews, thereby exercising greater interviewer control than may be possible in a personal interview survey.

**Questions Asked**    Questionnaire design is another very important and difficult part of survey design. A good survey design also includes testing of the questionnaire by interviewing a part (subsample) of the sample. Ambiguous or confusing questions can then be weeded out. Pretesting is probably the only way to ensure that answers to questions asked will lead to the desired responses. Here again, the telephone survey is advantageous since the survey director can get quick feedback from interviewers on the quality of the questionnaire.

**Timing**    The time when a survey is to be conducted should be considered beforehand so that the results are not unduly influenced. The timing of a survey on the viewing of public television, for example, is important. Should the survey be run in the fall when the new season for the commercial networks begins or in the spring when reruns are being shown on commercial stations? Certainly the time that is chosen will influence the results obtained.

An interesting aspect of the timing of a survey is that the timing can be used by the pollster to affect results. There are indications that polls on politically sensitive issues might be timed to benefit (or hurt) a politician! The commercial television networks apparently know when Nielsen is running a diary survey on viewing. Special programs are then carried during the survey week. This is why timing can be such an important design consideration as well as a strong influence on our interpretation of the survey results.

**Sample Size**    The size of the sample is an important consideration in any survey. A determination of the size sample to be drawn depends on cost and time limitations. The method of contact will also affect our choice of the sample size. At a fixed cost, more people can be interviewed by telephone than by personal contact.

Within the constraints of time and cost we should use the largest possible sample size. Recall that the standard error of an inferential statistic depends on the sample size (error decreases with an increase in sample size). The error of estimates must therefore be taken into account. If accurate estimates (low standard error) are absolutely necessary, a large sample size must be selected.

With regard to sample size note the following: the sample size must be adequate to each area that is to be used for statistical inference.

For example, suppose that for a statewide survey we are interested in a particular city in the state as well as the entire state. A sufficiently large sample size must be taken from that city. This will necessarily require a larger statewide sample. The more areas within the state on which we desire information (called *domains of interest*), the greater the demands for an increased statewide sample.

Other demands on sample size represent precautions against inadequate frames. For example, some numbers in a telephone directory are no longer in service. Thus, an increased sample size is required so that a certain number of respondents can be contacted. Or suppose we are contacting registered and unregistered adults in a pre-election survey. If, let's say, 60% of eligible voters are registered, only about 60 out of 100 people contacted are likely to be registered. If a person is unregistered his or her opinion on the candidate is not important as far as the survey is concerned. Thus, we need a large sample to ensure that a large enough number of registered voters are contacted.

Recall that the standard error of the sample proportion is the standard deviation of the sample divided by the square root of the sample size. From Appendix B, table 1, we observe that the standard deviation of the sample cannot exceed .5. The standard error can therefore be (over) estimated by .5 divided by the square root of the sample size. This gives us a pretty good estimate of the standard error in advance of sampling. The rule-of-thumb shown in table 1 of this chapter should be helpful here. The standard errors equal .5 divided by the square root of the sample size. Ninety-five percent confidence intervals for the population proportion are the sample proportion plus and minus two standard errors as listed in the table.

Recall from our discussions of the 1936 *Literary Digest* straw vote that sample size alone does not compensate for an inadequate sample design. In fact, no subgroup of the 8 influences mentioned in this section will themselves assure adequate survey design. All 8 influences must be considered.

**TABLE 1    Sample Sizes and the Associated Standard Errors for Domains of Interest**

| Sample Size (In a Domain of Interest) | Standard Error (of the Sample Proportion) |
|---|---|
| 100 | 5% |
| 400 | 2.5% |
| 900 | 1.65% |
| 1600 | 1.25% |

**Sample design**    Sample design is primarily concerned with increasing the precision of the estimates, that is, decreasing the standard error of estimates. For a fixed sample size we would seek the most precise estimates for a fixed cost or a set level of precision for the least cost. Recall from Discussion 5, chapter 5 that cluster sampling can increase error to a level determined by the number of clusters chosen rather than by the number of elements chosen. This decrease in precision is balanced with a decrease in the cost of running the survey. The cost-precision trade-off will not be discussed in detail here.

In the design of the sample we are concerned with ensuring that the sample is representative of the population of interest. But representative with respect to what — age, sex, marital status, height, weight? The sample must be representative according to the factors that will influence the response. Ideally, we should stratify (see the section on other sampling procedures in chapter 5) the population according to these factors and be sure that the sample is representative with respect to them. At least the sample should be checked after the interviews have been completed to verify that the sample is indeed representative with respect to these important factors. Such poststratification would necessarily require weighting responses. (A procedure for weighting responses was discussed in chapter 5.)

Let us consider some examples of designing a sample that is representative of a population.

Consider a survey for assessing the extent of the viewing of public television in a state. We must know the factors that influence the viewing of public television. Studies have found that the more highly educated segment of a population is more likely to watch public television. We would therefore want the sample to properly represent educational levels of people in the state.

With respect to a factor like education, stratification is impossible. There is usually no way to divide a population into educational groups (for example, last year of school completed was $K$–8, 9–12, 12+) and then to select within each group. We would, however, wish to ask people to which educational category they belong. We would then verify that the percentage of the sample in each category is the same as the population percentage (assuming, of course, that accurate population figures exist and that accurate information can be obtained from the respondent). If sample percentages in particular categories deviate greatly from population figures, responses in these categories should be weighted by the (population) sizes of the categories. These weights are relative to the entire population, however, so they are probably only obtainable from census records. In addition these records quickly become outdated, causing the weighting procedure to be of questionable value.

Other factors influencing the viewing of public television include income (which is in turn related to education) and the presence of children in the home. Children's shows like "Sesame Street" are very popular public television programs. These factors should also be considered when designing a sample.

Consider, once again, the preelection survey. Geographic stratification is often used in such polls. People living in a particular geographic area tend to vote in blocks. If these areas can be used as strata, a better sample design will result. Census tracting is a type of geographic stratification that would be helpful.

So, when designing the sample, we should consider the factors that influence certain responses. The design should ensure that the sample is representative of the population with respect to these factors.

Interpretation of survey results should be influenced by the considerations required for proper survey design. Designing a survey and judging a survey are closely interrelated. Basically, if you want a good survey, you must know how to judge a survey and how to design a survey that gives you results you can accept. If the results will be worthless, you should not waste your time and money.

## Design Considerations for Retrospective Studies and Prospective Studies

An important design consideration for retrospective surveys is the selection of controls. Recall from the last chapter that a control, or nondiseased, group must be selected for comparison with the diseased group. We discussed one problem associated with the selection of the control group. Hospital controls are often chosen because of the convenience of selection. Problems of inference to the general population arise as hospital patients are not likely to be representative of the general population.

A retrospective study can be designed to alleviate the problem of using hospital controls. Neighborhood controls could, for example, be selected. Let us look at an example. One of the more unusual outcomes of studies on the effects of smoking on health is that Parkinson's disease (shaking palsy) patients have been found to smoke less than hospital controls. Is this because hospital controls tend to be heavier smokers than the general public? (Parkinsonians are likely to be outpatients so that they are more like the general public in their smoking habits.) Another explanation for the difference is that nicotine acts to block the development of Parkinson's disease. A study using neighborhood controls needs to be run to help explain this unusual relationship between smoking and a disease. Neighborhood controls will be like the Parkinsonians in socioeconomic status. If Parkinsonians smoke less than the neighborhood controls, this would be strong evidence of the hypothesis that nicotine blocks the development of Parkinson's disease.

In chapter 7 we discussed age and sex matching of people in diseased and control groups. In this procedure a diseased person is identified, and someone of the same sex and age (within 5 years) is selected and placed in the control group. The diseased and control groups do not then vary with regard to age and sex. This is desirable as the factors of age and sex might themselves be antecedent factors. As we've discussed, if the diseased and control groups vary by more than one antecedent factor, the results of a retrospective study may not be conclusive. Note that age is likely to be an antecedent factor for most diseases as older people are more likely to develop disease. Many diseases have different incidence rates for men and women. It is likely then that sex is predictive of disease.

Problems can arise when controls are not age-matched to diseased people. Hospital controls often come from the hospital staff, which is usually younger as a group than a group of diseased individuals. Suppose then that smoking characteristics are compared between the groups. If pack years are used, a difference in smoking will likely exist. Recall that the pack-years figure depends on the number of years a person has smoked. The diseased group will show up to be greater smokers as older people will have smoked for a longer period of time. Hence, the difference between the average number of pack years among the groups is likely a reflection of the age difference, not a reflection of difference in smoke consumption. Age matching must be used when comparing any two groups with regard to pack years.

Because of the size required for a prospective study an entire population is usually studied. Studies of all physicians in Great Britain, all registered nurses in the United States, all United States veterans, and everyone in Framingham, Massachusetts are examples. Each of these groups represents a population. Investigators strive for complete participation. Such a study is a census with only nonparticipants in a population not being studied.

One prospective study has certain interesting design characteristics worth noting. The Royal College of General Practitioners in Great Britain has been observing women to see the effects of oral contraceptive use on the circulatory system.[2] This study was referred to in the section on observational research in chapter 7. Women who use birth control pills were matched with nonusers. (Each group contains 23,000 women.) These two groups have been observed over an extended period. The matching of people with and without a particular antecedent factor is a new approach to a prospective study. Matching is too often restricted to retrospective studies. Hopefully, we'll be seeing more of this type of design for a prospective study.

## Designing Experiments

In contrast to a sample survey, in an experiment the population being studied is created for the purpose of the investigation.

> The word *experiment* refers to the study of a population that is created for the purpose of investigation.

For example, when testing whether a drug is carcinogenic, an experimenter creates a laboratory situation that simulates the use of the drug. An examination of this artificial situation tells the investigator about the nature of the drug's behavior. The investigator can then determine if the drug should be marketed, and the decision to market may lead to the natural development of the population the experimentor tried to simulate. As another example, consider an experiment to study a certain type of fertilizer. The use of the fertilizer on an experimental test-plot simulates the population that would exist if farmers did indeed use the fertilizer on their crops.

### *Important Considerations in Experiment Design*

Like survey design, good design of an experiment requires that the factors that might influence a particular response be known and properly handled. The idea in experiment design is that all but one factor (drug, fertilizer, chemical, etc.) are the same in, let's say, two groups — that is, the two groups differ only by the treatment or factor we wish to test. Hence, any difference between the two groups can be attributed to this key factor.

In experimentation researchers must be sure that the situation created will ensure answers to the questions posed. A major concern of researchers is that their own biases and the biases of their associates do not improperly influence the population they must create. This

concern of the researchers is not too far removed from the pollsters' concern that their sample properly represent the population of interest.

**Control**   The term "control" in controlled experiments is used to describe a group not receiving treatment; the control group may represent an existing population. The population that the researcher creates will be compared to this control group. More than one treatment group can be compared to a control. The control group should vary from the treated group(s) only with regard to the factor(s) or treatment(s) of interest. In this way any difference observed between the groups can be attributed to the treatment(s).

For example, let us consider the mid-1970s test of the effect of saccharin on rats. During this experiment rats were fed a diet consisting of 5% saccharin for their entire lives. Bladder tumors were counted. Before the effect of this level of saccharin on bladder tumor development could be evaluated, however, researchers needed to know how many of the rats would have developed bladder tumors without saccharin in their diet. Control rats were used for this purpose. Two groups of rats were housed and fed identically, except one group had saccharin added to its food. Any difference in the number of bladder tumors between the two groups could then be attributed to saccharin.

Not all experiments involve the use of a control. Sometimes treatments are compared to each other but not to a control. In this situation many populations are created for the purpose of comparison.

**Measurements**   When designing any experiment we must determine the measurement that is to be made. This necessarily involves the determination of a clear definition of the factors of interest. From our discussions in chapter 2 it should be clear that an improper choice here could weaken the conclusions of the most carefully designed experiment. For example, in the field study for testing the Salk polio vaccine, the measurement taken could be a count of deaths, cases of paralysis, or all cases of polio including mild (flulike) cases. (What would be the advantages and disadvantages of each choice in this case?) Also, recall from the last chapter the problem involved with measuring the levels of smoke consumption (cigarettes, cigars, and pipes).

Measurement problems vary among areas of statistical investigations as well as for the different problems within an area. In agricultural research measurements would be continuous. For example, weight gain of hogs fed a special ration and yield of a crop using different fertilizers are continuous measurements. Yet, some subjective measurements in agricultural research could require work with ordinal or nominal measurements. For example, the quality of a tobacco leaf might be judged by classification according to aroma, which would be an ordinal scale, with the classes "below average," "average," "above average."

In the behavioral and social sciences measurement problems can be very profound. Here an investigator is often interested in subjective measurements such as a person's feelings about his or her relationships with peers or the needs that a person perceives as not being met by different governmental agencies. A whole experiment could hinge on the researcher's ability to find an acceptable measuring device or instrument.

Other important experiment design considerations include sample size and precision of estimates. In the case of dichotomous responses the sample size necessary for a particular level of precision can be determined as in the preceding section: the most precise design is the design that provides the smallest inferential error. We will not discuss this aspect of experiment design. (See the Further Readings section for more information on this topic.)

**Randomization** Since some factors that might influence response are difficult if not impossible to control, subjects are often assigned to the treatment (including the control) groups in a random manner. The conscious or subconscious biases of the technicians involved in the experiment should not be allowed to influence the response of interest. Randomization guards against:

1. Lack of balanced judgment entering into the construction of treatment groups

2. Overcompensation by a researcher to protect against a bias, creating a bias in the opposite direction

3. Criticism of experiment results[3]

Recall that self-selection was a disadvantage of a prospective study. People choose which group they will be in — they are not randomly assigned to groups with and without the antecedent factor. In general, a lack of randomization makes the results of a survey or experiment difficult to interpret.

Randomization is necessarily used when experimental units are alike, or homogeneous. *Experimental units* are the smallest units that receive treatment in an experiment. Examples of an experimental unit are an animal, a plot of land, and a tissue culture. Homogeneous experimental units would be, let's say, mice of the same age, size, sex, and species.

A group of people could be a homogeneous group for an experiment if they are alike with regard to the factors likely to influence response. For example, in many human studies on disease, age is a factor that often influences response. (Response in this case is the occurrence of a disease.) People of the same age would be a homogeneous group of experimental units as long as these people did not differ by any other factor(s) that might influence response. (More than one factor may be known to influence a response.)

In the saccharin example previously mentioned the rat is the experimental unit. The rats are bred to be a pure strain; that is, the rats are homogeneous experimental units. Rats are then randomly assigned to control and treatment (saccharin) groups.

**Other Safeguards**    Sometimes experimental units may not be homogeneous. One would not want to completely randomize in this case. For example, when randomly placing people into treatment groups for an experiment concerning a disease, we might get all of the older people in one group. If age is related to the occurrence of a disease, we may get a biased response from the treatment group made up of older people. If age (or any other factor) is likely to influence the response, we should block or stratify according to this factor and then randomize within the block. In the situation in which age is an important factor, we would randomly assign people of, let's say, 20 to 29 years of age to the different treatment groups.

---

A *block* is a part of a created population. Experimental units within blocks should be homogeneous with respect to the factors that might influence response. (Observe that a block is analogous to a stratum of an existing population.)

---

Other precautions against undesired influences on response are provided by placebos and double-blind procedures.

---

A *placebo* is a treatment like the treatment of interest in appearance but inactive, except for psychological effects, as far as the response of interest is concerned.

---

The idea behind the use of a placebo may be illustrated by an example. In a test of the effectiveness of a headache remedy, people getting a pill may think they got relief. To measure this effect, a control group may be given a sugar pill or placebo. Because of the use of placebos for the control group, the differences between the control and treated groups are due to the effects of the treatment over and above any psychological effects.

This idea may also be used in experiments involving animals other than humans. The act of administering treatment may cause a response. In order to smoke an animal for an experiment in smoking and health research, an animal may have to be restrained. The restraining of the animal may affect, for example, eating habits so that a look at the relationship between smoking and nutrition would be an example in which response would be influenced by the restraining process. In such cases a group of animals may be *sham-smoked,* or restrained, but not smoked. A comparison of a control group, a sham-smoked group, and a smoked group would then be desirable so that the effect of smoking (alone) can be properly evaluated.

Some measurements are subjectively determined so that a technician may think he or she sees a response because the people or objects being examined were treated. A double-blind design is helpful here.

> In a *double-blind* experiment neither the technician nor the person being tested knows who was treated.

## Examples

Let us look at the experiments discussed in the section on experimentation in chapter 7 so that we might get a better idea of the use of randomization, a placebo, and double-blinding as well as the use of a control group. Recall our discussion of the 1747 experiment conducted by James Lind.[4] This experiment established that scurvy is caused by a lack of Vitamin C, which thereby provided a cure. Lind experimented on 10 men with scurvy. The cases of scurvy were similar in severity. The 10 men were divided into groups of two and tended the same way, except for one of 5 different treatments administered to each group of two:

> They lay together in one place, being a proper apartment for the sick in the fore-hold of a ship, and had one diet common to all, vis. water-gruel sweetened with sugar in the morning; fresh mutton-broth often times for dinner; at other times puddings, boiled biscuit with sugar, etc. and for supper, barley and raisins, rice and currants, sago and wine, or the like.[5]

Lind should have been sure that the 10 men were *randomly* assigned to the 5 treatment groups. As we see these men formed a group of homogeneous experimental units. Hence, randomization is called for. Randomization would help control against Lind's own bias since he may have consciously or subconsciously assigned the less severe cases of scurvy to the treatment he felt most likely to be successful. Lind did not apparently use random assignments.

A control group for Lind's experiment would have been two scurvy victims housed as the others but not given a special treatment. Lind did not use a control group.

Walter Reed's (1901) experiment concerning the cause of yellow fever is an example of controlled experimentation on human subjects.[6] Reed's Camp Lazear (near Quemados, Cuba) experiments involved three groups of human volunteers. One group, a treatment group of 7 men, lived in a poorly ventilated "Infected Clothing Building," which received little, if any, sunlight. These men literally slept in the very beds and garments of yellow fever patients. (These men were not bitten by any mosquitos.) The other two groups of men were housed in a building partitioned into two rooms by a wire mesh through which mosquitos could not penetrate. All things were the same in both rooms except that mosquitos were released in one room. These mosquitos had fed on yellow fever patients.

Of the 13 volunteers bitten by mosquitos, 10 contracted yellow fever. Of the three not contracting yellow fever from mosquito bites, one must have been immune as not even injections of the blood of yellow fever patients gave him the disease. The other two who did not get yellow fever from mosquito bites did get the disease from blood injections. It was theorized by Reed that one of these two men was bitten by mosquitos too young to have developed the disease. (Mosquitos can transmit the disease only after a period of incubation following the biting of an infected person. This period was reported by Reed to be at least 11 days. What effect might this have had on the experiment design?) The second of the two who did not contract the disease from the mosquito was bitten by mosquitos who bit a yellow fever patient only 8 hours after onset of the fever's symptoms. The microorganism responsible for the illness was probably not yet in the victim's bloodstream, and the mosquito itself had not assimilated the microorganism. No one contracted the disease by living in the control side of the building or in the Infected Clothing Building.

Four of the 5 volunteers who were injected with the blood of yellow fever patients got yellow fever. Two of the 5 men were from the 13 subjected to mosquito bites but did not contract yellow fever; the other 3 were not in the group of 13 subjected to insect bites. This provided evidence of how the mosquito could assimilate the microorganism, which it then passed to other people it bit.

The third human-subject experiment mentioned in chapter 7 involved a field study on the effectiveness of the Salk vaccine against polio.[7] Factors influencing the occurrence of polio needed to be considered in the design of this experiment. First, the rate of occurrence of polio varied greatly from year to year so that establishing a control and treated group was important to ensure that any reduction in the incidence of the disease for the vaccinated group over and above the control group could be properly judged. The treated and control groups would have to show a significant difference in rates of occurrence of polio for the particular year of the test — 1954.

A large sample size (about 750,000 for the experiment being discussed and over 1 million additional children for a second experiment run concurrently) was required since the incidence rate of polio was low (about 71 in 100,000 cases in the year of the study).

Since not all cases of polio are paralytic, the diagnosis of the attending physician was an important design consideration. Whether a flulike sickness was indeed polio was a somewhat subjective determination. A double-blind procedure was used. Neither the child (first, second, or third graders) nor the doctor knew who was vaccinated by the Salk vaccine or who was vaccinated by a salt solution (the placebo). The doctor's judgment on whether the disease was contracted by a child would not therefore be consciously or subconsciously influenced by the knowledge of whether the child had been innoculated by the Salk vaccine.

Randomization was also used in this field study. Vials containing the Salk vaccine and the placebo were numbered and shipped together to test areas. A child was inoculated with the contents of a bottle. The number on the bottle and the child's name were recorded. Only upon completion of the experiment were the numbers decoded and a determination made concerning which children received the Salk vaccine.

As a last example of experiment design, let us look at a nutrition experiment on rats. (Nutritional requirements of rats are not too dissimilar from that of humans.) Once weaned, rats are put on a standard laboratory diet for about a week. In an experiment on breakfast cereals one group of rats was fed nothing but water and a particular cereal after being weaned. Three other groups, the control rats, were fed only milk, only eggs, or only a regular laboratory diet. Rats were fed this way for 23 weeks, unless a weight loss was observed. Rats who lost weight were put on the laboratory diet to determine whether this weight loss was due to their diet or to some other cause. Using weight gains of rats on the laboratory diet as 100, eggs rated 82 and milk rated 55. Cereals were similarly rated.

Good experiment design requires that the rats be randomly assigned to treatment (diet) groups and housed in a similar manner. The only differences between rats in the different groups should be the dietary differences.

Note that laboratory animals are bred so they are homogeneous. A major advantage of animal experimentation is the fact that genetically homogeneous experimental units exist. Couple this with randomization and a good experiment can be designed.

It may be easier to conduct experiments on laboratory animals, but a question should be raised here: Does reaction of a laboratory animal to a treatment necessarily give an accurate indication of how humans might react to this treatment? For example, will humans necessarily develop cancer from a chemical that produces cancer in rats? Whatever the answer to these questions, in many situations there is no other practical or ethical way that better information can be obtained.

# General Considerations in Design

We have discussed certain aspects of designing a sample survey and an experiment. Recall from the beginning of this chapter that these types of investigations differ by the population being studied. In a sample survey we study an existing population. In an experimental situation a population is created to be studied. Let us now look at certain design considerations for the study of any population, whether it exists or is created.

The study of a population is a three-phase investigation: the planning phase, the design phase, and the analysis phase.

## *Planning Phase*

The planning phase begins with a statement of the objective(s) of the study. The investigator must state very precisely what he or she hopes to accomplish through the study of a population. This is an important part of any study. An investigator will bring data to a statistical consultant and ask what is in the numbers. One of the first questions the statistician will ask is the reason the data were collected in the first place. Hopefully, this will not be the first time that the experimenter considered the question. When designing a study of a population, an investigator should first state the objectives of the study.

Another part of the planning phase is a determination of the numbers to be collected. As discussed in chapter 2 and earlier in this chapter an experimenter could be wasting his or her time if definitions are not clearly stated and measurements are not properly taken. In certain agricultural experiments in which physical measurements such as yield are taken this part of the planning phase may not be as crucial as in investigations in the social and behavioral sciences. As discussed before the definition-measurement problems in the social and behavioral sciences can be a most difficult design consideration.

Another part of the planning phase of an investigation is the specification of the population of interest. For a sample survey we must determine the population of study, or target population. In experimentation investigators must determine what population or populations they wish to create. This involves a determination of the treatments to be given experimental units.

### Design Phase

The second phase of a population study is the design phase. In this phase an investigator needs to first determine the observational or experimental units. The observational unit is the smallest unit that is observed. For the sample survey this unit is likely to be a person. The experimental unit is the smallest unit to which a treatment is applied when an investigator is creating a population. Experimental units can be laboratory animals, plots of land, etc. Experimental units correspond to observational units in a sample survey.

Another aspect of the design phase of a population investigation is a determination of the number of observations to be taken. How large should a sample be in a survey, or how many experimental units should be treated in an experiment? An investigator must work within certain cost and physical constraints. The idea is, of course, to get the best data within the framework set by these constraints.

Observations will vary from one population element to another, whether the population exists or is created. The better able we are to explain this variation, the better able we will be to design an investigation that is as accurate as possible. In experimentation we try to control potential sources of variation or effects on response other than treatment. In sample survey design we might stratify according to the variables that affect the response of interest. "Blocking" is a term used to describe a similar device used in experimentation.

Randomization is another important part of the design phase. We must decide how to randomly select observations from an existing population. Likewise, we need to determine the order in which experimental units are given treatments. The word "design" is used to describe the *order* in which experimental units receive treatment. A completely randomized design is one in which experimental units receive treatment in a completely random manner. For example, to decide which mouse in a (homogeneous) group of mice will receive a particular dose of a drug, a mouse is selected at random. A completely randomized block design refers to random assignment of treatment(s) to experimental units within a block.

The term "experiment" is often more subtly and more properly used to refer to the treatments to be given experimental units — that is, an experiment is the population(s) created for study. The word "design" more properly refers to the *order* in which the treatments are given experimental units.

The last part of the design phase requires an outline of results. It is desirable to plan the data anlysis before the data are collected so that any potential problems might be found beforehand. I have seen data that cannot be analyzed because this preanalysis part of the design phase was skipped.

### Analysis Phase

The final phase of the design of an investigation of a population is the analysis phase. At this point the data are collected by either sampling an existing population or observing a created population. Too often an investigator will start at this point in the study. Problems can clearly arise in such instances.

After the data are collected, they must be analyzed and the results interpreted. The analysis may not be straightforward even with careful work in the design phase. Inaccurate data can come from the most carefully designed investigation because technicians or interviewers make mistakes, animals die from natural causes, test-plots of land are flooded, etc. The collection, processing, analysis, and interpretation of data can be a very involved process. We have seen this from our discussions of the interpretation of statistics, which is itself very involved.

### The Role of the Statistician

The investigation of a population, existing or created, is a complicated process requiring the cooperation of statistician and investigator. It is this role of the statistician that best describes what the discipline of statistics is.

The role of the statistician in assisting the investigator can be seen to be twofold. First, a statistician must be aware of the tools available to the investigator so that the most accurate study of a population can be carried out in the most efficient manner within the cost and physical constraints imposed on the investigation. Second, the statistician and investigator must be able to communicate with each other. This is just as important as familiarity with the skills of statistics since a statistician who knows everything there is to know about the tools of the trade is of no use to an investigator if the statistician cannot convince the investigator what should be done as well as what was done in a population study.

Statistics is therefore a set of tools for use in the study of a population as well as a language. In this chapter we have seen the discipline of statistics as a set of tools that aid an investigator in the study of a population. However, our efforts in this book have been, for the most part, to teach you about the language of statistics.

# Summary

We have taken another look at the influences on the interpretation of survey results: the background or subjective influences (population sampled, method of contact, response rate, wording of questions, and timing) as well as the influences of sample size and sample design.

The size of the sample indicates the accuracy of inferential statistics. The sample should be designed so that, as far as possible, it represents the population with regard to the factors that might influence response. If, for example, we are questioning adults, we would like to know the factors that might affect their answers. We wish to be sure that the sample represents adults in the population with respect to these factors.

When designing an experiment we must once again determine the factors that might influence response. To ensure that the biases of the experimenter and the uncontrolled factors affect the different treatment groups equally, homogeneous experimental units should be randomly assigned to treatment groups. A placebo should be given a group of experimental units if the act of administering a treatment might influence response.

To ensure that responses are not improperly judged, a double-blind design is often used. Since technicians might be more likely to see a response if they know whether the experimental unit was treated, the technician is not told which units were treated. In the case of human experiments, the subjects will not know whether they received treatment (hence, the word "double").

The study of a population, whether existing or created, is made up of three phases. During the planning phase the objectives of the study are specified. The study is set up during the design phase. Our discussions in this chapter have been directed toward this phase of a study. The final phase is the analysis and interpretation phase. Our discussions throughout this book have been directed toward interpretation of the results of a study rather than analysis of the data generated.

Seeing how a statistician acts as a consultant in the investigation of a population gave us an opportunity to view statistics as a discipline. Statistics is the tool required of a researcher in the study of a population. Statistics is also a language. The language of statistics has become extensively used by the media as politicians and advertisers rely on this language more and more. A large part of our discussions in this book has been geared to understanding statistics as a language, rather than understanding statistics as a research tool in the study of a population.

# No Comment

> ■ *Too often, in the pursuit of progress and profit, our health and the health of our environment, where we work and where we live, are secondary matters. Too often, our environment is the laboratory — and we are the guinea pigs."*
>
> *(Morton Dean, CBS news correspondent,*
> *"The Case of the Plastic Peril,"* CBS Reports, *19 October 1974)*

# Exercises

1.  In January of 1976 the Food and Drug Administration took Red Dye # 2 (Red 2) from the United States market. This decision was based on the following "experiment":[8] Five hundred rats were (randomly) placed into 4 groups. Each group was given a different dose of Red 2. This experiment was faulty because:
    a.  The study was left unsupervised for a long period of time after a scientist was transferred.
    b.  An undetermined number of rats were mixed or shifted between treatment (dose) groups.
    c.  Dead mice were not quickly removed from their cages so that cause of death could be determined. (It is always important to know if death resulted from treatment.)

    Yet, the results of this experiment led to the discontinuance of Red 2 as a food coloring. The 96 surviving mice were put into two groups, "high dose" (3% of their diet was Red 2) and "all others" (a low-dose group). It was observed that there was a significant increase in the number of malignant tumors in the female rats fed the high dose. When this fact was made public, Red 2 was pressured off the market.

    How do such "statistics" fit into the decision-making process discussed in chapter 4? A different definition of when a tumor is "benign" or "malignant" would change the claim of a significant difference between the "high dose" and "all others" groups. What effect does this have on results? What violations of proper experimental procedure were committed in this experiment?

2. The Salk polio vaccine was a dead virus vaccine. The polio virus is reduced in strength by formaldehyde to a level that reaction to the virus is not serious but causes the body to produce the desired antibodies.[9] A batch of this vaccine was not properly prepared and actually gave some children polio.[10] In one case 4 out of 30 monkeys injected by a batch of the vaccine developed mild to severe brain damage. Yet this batch of the vaccine was put on the market. Describe this test of hypothesis. What of Type-I and Type-II errors? Should the size of the test (level of significance) be large or small?

3. Consider the following design of an experiment to test different paints.
   a. A couple of hundred houses were selected nationally, with consideration given to temperature range, sunshine, and mildew.
   b. Houses were owned by average people. The houses were stucco, brick, wood shingle, plywood, and wood clapboard.
   c. These houses were painted with 6 brands of paint as indicated in figure 1.

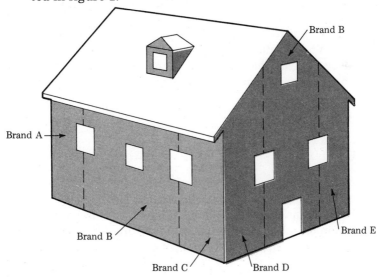

**FIGURE 1**    **House-Painting Experiment**

    d.  Comparisons were made between one particular paint and the 5 other brand-name paints. Paints put on the same side of a house were compared. Brands B and D, D and E, and/or B and E could be compared on the side of the house in figure 1 with a door.

    e.  In the wooden clapboard houses where paint cracking occurred, a competitive paint cracked first. The results favored one brand 250 to 15; 7,000 comparisons were made.

        Does this appear to be a well-designed experiment? Any background influence of importance? How about the way data are reported? Any data not reported?

4.    In chapter 6 two advertising claims of significance were mentioned: an analgesic rub improved grip strength significantly, and a product containing caffeine helped drivers maintain alertness (significantly) better. How might the companies involved design credible experiments that could be the basis of these advertising claims? Consider who should be tested and how the testing should be carried out.

5.    I talked to a student who wanted to know whether a particular survey was well designed. This student was working for a state agency for the summer. The agency proposed interviewing 5,500 people (an expensive undertaking). These 5,500 people were to be selected by randomly drawing 110 equal (land) area clusters in the state and then taking 50 interviews in each cluster. (If a cluster was too sparsely populated, adjacent areas were used.)

        What would you tell the student about this design? Consider, for example, that in a county with one third of the state's population, 150 interviews were called for with this design.

# For Discussion

1.    Many products (for example, Red Dye # 2, cyclamates, and vinyl chloride) have been removed from the market. Discuss this decision-making process, recalling our discussion of experimentation on laboratory animals and of the influences on the decisions that people make (chapter 4). Are there any political considerations here?

2.    Discuss the execution of a survey. What effect will the execution have on the credibility of the survey? Relate this to the design considerations discussed in this chapter.

3. Discuss the time and dosage problems of animal experimentation. For example, in Exercise 1 3% of the diet of the mice in the "high dose" groups was Red 2. Time problems are involved in experiments with animals since a person will consume a product at low levels for many years. What would be the equivalent dosage for a rat whose life expectancy is three years?

4. Discuss design considerations of a sample survey of your state. You pick the subject of the inquiries. (How might your determination of subject affect the survey design?)

5. Discuss the need to recall people in a survey. Can this be tied into our discussion of quota and stratified random sampling techniques in chapter 5?

6. Suppose that a retrospective study on the incidence of heart attack and the rising use of oral contraceptives is to be designed. What control group would you recommend? Would you match controls and heart attack victims? According to which variables would you match?

7. Discuss matching groups of people with and without a particular antecedent factor in a prospective study. Does this solve the problem of self-selection? What are the advantages — that is, are the results of such a study more credible than the results of prospective studies without matching?

8. A design problem of which we should be aware exists in the way that a treatment is administered to experimental units. For example, a compound, called 4-MMPD, found in certain hair dyes has been found to be carcinogenic; that is, the compound produces cancer in laboratory animals. But in the experiments the compound was administered to mice by ingestion or injection. If this compound is to cause cancer in humans, it would have to be absorbed through the scalp. Discuss potential problems of inference in this case. How might a better experiment be designed?

9. Discuss inference to a nonexistent population that is created in an experiment. How restrictive is the inference?

# Further Readings

Some of the books mentioned at the end of chapter 5 may be an aid in learning survey design. Another book of interest in this regard is Des Raj's *The Design of Sample Surveys* (New York: McGraw-Hill, 1972).

For a readable discussion of both sample survey and experimental design see W. Federer's *Statistics and Society: Data Collection and Interpretation* (New York: Marcel Dekker, 1973).

# Notes

1. M. Wheeler, *Lies, Damn Lies, and Statistics* (New York: Liveright, 1976), chapter 1.
2. Royal College of General Practitioners, *The Lancet* 2 (8 October 1977): 727–733.
3. Sir A. Bradford Hill, *Statistical Methods in Clinical and Preventive Medicine* (New York: Oxford University Press, 1962), p. 35.
4. *Ibid*, p. 4.
5. *Ibid*, pp. 4–5.
6. W. Reed, "The Propogation of Yellow Fever: Observations Based on Recent Researches," *Medical Record* 60 (1901): 201–209.
7. P. Meier, "The Biggest Public Health Experiment Ever: The 1954 Field Trial of the Salk Poliomyelites Vaccine," *Statistics: A Guide to the Unknown*, ed., J. M. Tanur (San Francisco: Holden-Day, 1972), pp. 2–13.
8. P. M. Boffey, "Color Additives: Botched Experiment Leads to Banning of Red Dye No. 2," *Science* 191 (1976): 450–451.
9. P. Meier, "The Biggest Public Health Experiment Ever: The 1954 Field Trial of the Salk Poliomyelites Vaccine," *Statistics: A Guide to the Unknown*, ed. J. M. Tanur, pp. 2–13.
10. N. Wade, "Division of Biologies Standards: Reaping the Whirlwind," *Science* 180 (1973): 162–164.

# Correlation and Regression 9

THE RELATIONSHIP BETWEEN DIFFERENT CHARACTERISTICS is often of interest. For example, what is the relationship between the summer weather as measured by temperature and the amount of electricity a particular utility company must be able to furnish its customers? If a relationship between temperature and demand does exist, we wonder if we could predict demand using temperature. Then, if the temperature is expected to increase, the utility company can make arrangements in anticipation of increased demand.

This is the type of problem we will look at in this chapter. We will restrict our attention to two variables, that is, we will work with data pairs. In practice many measurements might be of interest. For example, time of day, time of year, and other weather variables besides temperature may be of aid in the prediction of electrical demand. A less involved look at this problem using only temperature and demand will, however, help us understand the basic concepts of correlation and regression.

We will also discuss a simplified approach to the type of data that a university would look at to determine whether an applicant should be admitted to a particular program of study. Our example will consider admittance to graduate school. In this case the data pairs of interest will be the undergraduate grade point average (GPA) and graduate school GPA of students at the university in question. This model is simplified since, in reality, a university would consider more than just these two measurements; also included might be quality of the undergraduate institution and/or the score on some standardized admissions examination such as the Graduate Record Exam (GRE).

We will be considering data pairs, or two measurements taken from each element in a population. We will be looking at the relationship between the two measurements of interest. For example, we might wonder how a person's height is related to his or her weight. We might suspect that taller people tend to weigh more. Similarly, we would expect higher summer temperatures to mean greater demand for electricity needed to run air conditioners.

We will take a twofold approach to the study of the relationship between two measurements. First, we will describe the relationship between the two measurements. A graph (the scatter diagram) and a statistic (a correlation coefficient) will be used for this purpose. Second, we will investigate the prediction of one measurement using the other. In addition we will discuss how the prediction of one measurement from another can be carried out by using a line that we fit to the data pairs and we will consider an index of accuracy of the prediction.

We will also investigate testing whether two variables are related. The case when both measurements are continuous and the case when both observations are dichotomous will be considered.

Once again we caution against inferring a cause-and-effect relationship between responses when a high statistical relationship exists between the associated measurements.

## History

The predicted value of a measurement is often referred to as the *regressed value.* The prediction problem is then referred to as the *regression problem.*

Investigation of *correlation* (from "co-relation") and *regression* (or "reversion") was begun with Sir Francis Galton. Galton, whose IQ was estimated at 200, was a half cousin of Charles Darwin. Like his half cousin, Galton traveled extensively. He had a passion to count and measure — "heads, noses, eye color, breathing power, strength, reaction time, frequency of yawns, number of fidgets per minute among persons attending lectures."[1] It is no wonder that Galton was responsible for the application of statistics to so many areas of study outside the sciences.

In his *Memories* Galton writes:

As these lines are being written, the circumstances under which I first clearly grasped the important generalization that the laws of Heredity were solely concerned with deviations expressed in statistical units, are vividly recalled to my memory. It was in the grounds of Naworth Castle, where an invitation had been given to ramble freely. A temporary shower drove me to seek refuge in a reddish recess in the rock by the side of the pathway. There the idea flashed across me, and I forgot everything else for a moment in my great delight.[2]

Galton wrote thus about his first thoughts on the apparent regression of natural characteristics toward the mean.

How is it possible for a population to remain alike in its features as a whole, during many successive generations, if the average product of each couple resemble their parents? Their children are not alike, but vary: therefore some would be taller, some shorter than their average height; so among the issue of a gigantic couple there would be usually some children more gigantic still. But from what I could thus far find, parents had issue less exceptional than themselves.[3]

Galton's observations suggested a regression to mediocrity in height from generation to generation. Hence the term "regression." (Galton's interpretation was fallacious since he implied that people would tend to the same height. A better explanation would be that unusually tall people have an unusual genetic makeup that is not likely to be passed to the offspring. This is for unusually tall people. It would still be true that the taller of two people is likely to have the taller offspring.) The theory of correlation and regression was refined by Karl Pearson and others in the late nineteenth century.

Today the prediction problem is used in many varied areas of research including business and medical research. Using time as one of the measurements of interest businesses try to predict marketing trends in the future. In medical research a dose-response curve is extensively used. The response of experimental units is observed for varying doses of a drug. One important outcome of such research would be to find the optimal dose of a drug, that is, to find the dose of a drug that will give a desired response without undesired side effects. Let us take a look at this very important area of statistics.

## Graphic Display of Paired Data:
## The Scatter Diagram

Graphic representation of paired data is in the form of a *scatter diagram.* The horizontal scale represents one measurement, and the vertical scale represents the other measurement. A dot is placed above the point representing the measurement on the horizontal scale and across from the point representing the (paired) measurement on the vertical scale. As an example, let us look at a graph of the data pairs (2,3) and (1,2) in figure 1. To represent the pair (2,3) a dot is placed above the 2 on the horizontal scale and across from the 3 on the vertical scale. The pair (1,2) is represented in a similar manner.

We will be considering a more realistic example in some detail. Let us look at the area temperature and the electrical output of a

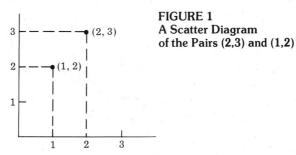

FIGURE 1
A Scatter Diagram
of the Pairs (2,3) and (1,2)

utility company. Our objective will be to find a way to predict demand on the company's system. This is important for a utility company since, for example, high demand would necessitate using extra generators or even buying additional electricity from a neighboring company.

Many bits of information would aid a company in its attempt to predict demand. For example, the company would want to know the time of year (summer demand due to air conditioning varies from demand at other times of the year), time of day (daytime demand when people are at work usually exceeds nighttime demand), as well as many different weather variables. We will restrict our example by looking at only one weather variable, temperature, and fixing (holding constant) the time variables. We will consider only 10:00 AM readings on weekdays in July. The temperature and corresponding demand (in megawatts) are given for the 22 weekdays in July 1971.

We begin our look at these data by describing the 22 pairs of numbers. Table 1 lists the data.

The temperature for 1 July is 25.0° C. with an output of 1,156 megawatts. The date 1 July is represented in the scatter diagram of these data in figure 2 by a dot above 25.0° on the horizontal (temperature) scale and across from 1,156 on the vertical (output) scale. The remaining 21 days are similarly represented by dots in the scatter diagram. (Some days are labeled in figure 2 so you can see how the diagram is constructed.)

Descriptive information may be extracted from the scatter diagram. We first observe that the dot representing 5 July is far from the dots representing the other days. Upon reflection we note that 5 July was a Monday, and since 4 July fell on a Sunday, 5 July was a holiday. The 5 July data pair, called an *outlier,* should be considered with Sunday data. We will therefore look at only 21 data pairs, excluding 5 July.

**TABLE 1    Temperature and Electrical Output Data**

| Day (July) | X (Temp) | Y Output | Day | X (Temp) | Y (Output) |
|---|---|---|---|---|---|
| 1 (Thurs.) | 25.0 (77)* | 1156** | 16 | 25.0 (77) | 1174 |
| 2 | 23.9 (75) | 1073 | 19 | 22.2 (72) | 1105 |
| 5 | 25.0 (77) | 889 | 20 | 21.7 (71) | 1066 |
| 6 | 22.2 (72) | 1045 | 21 | 22.8 (73) | 1062 |
| 7 | 26.1 (79) | 1155 | 22 | 25.6 (78) | 1078 |
| 8 | 27.2 (81) | 1213 | 23 | 25.6 (78) | 1126 |
| 9 | 29.4 (85) | 1273 | 26 | 26.1 (79) | 1179 |
| 12 | 23.9 (75) | 1130 | 27 | 22.8 (73) | 1068 |
| 13 | 25.0 (77) | 1136 | 28 | 23.9 (75) | 1063 |
| 14 | 23.9 (75) | 1111 | 29 | 21.7 (71) | 1081 |
| 15 | 25.0 (77) | 1121 | 30 | 20.6 (69) | 1061 |

*Temperature in degrees Celsius (Fahrenheit)
**Megawatts of power output

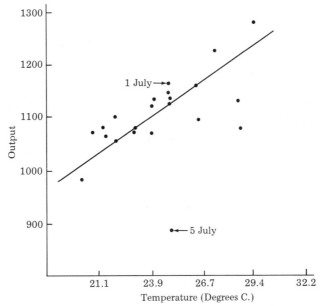

FIGURE 2    Scatter Diagram of Temperature and Output Data

Another bit of information obtained from the scatter diagram is the determination of whether the data is approximately linear (like a straight line).

> Generally speaking, if the scatter diagram tends to follow an elliptical (oval) pattern, the data is said to be *linear*. (As we will say later, the data fit a linear model.)

Examples of such data are given in figure 3. (We will discuss the correlation coefficient, which is indicated on the graphs, in a later section.) Note that the scatter diagrams in figure 3 range from circular patterns to elliptical patterns that follow very closely to a line. (A circular scatter diagram indicates a lack of association. The term "linear" is vacuous in this case since there is no relationship to describe.)

Data need not necessarily be linear. A scatter diagram of non-linear, curvilinear data is given in figure 4. Observe that the dots in figure 4 do not tend to follow an elliptical pattern: the pattern is *L*-shaped.
pattern: the pattern is *L*-shaped.

The broken line in figure 4 is a curve that seems to reflect the pattern of the dots in the scatter diagram — that is, the dots in the

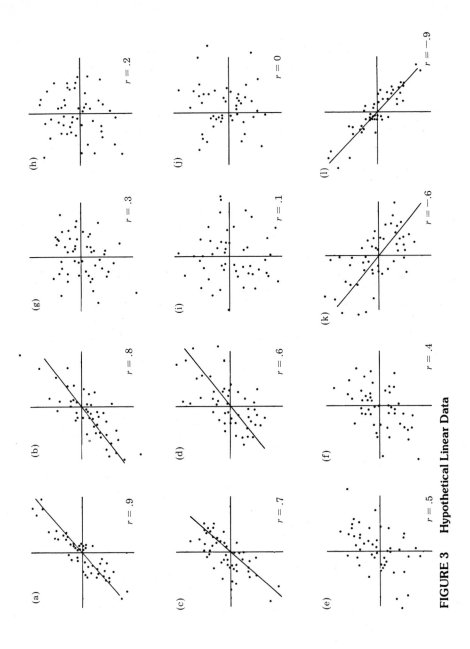

**FIGURE 3**   Hypothetical Linear Data

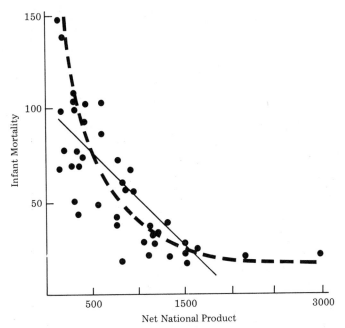

**FIGURE 4** **Scatter Diagram of Nonlinear Data**

Source: D. Heer, "Economic Development and Fertility," *Demography* 3(1966): 423–444.

scatter diagram vary about this curve. You can see that there is not a tendency for these data to vary about a line in any consistent manner. (See, for example, the line through the dots in the figure. The data do not tend to deviate about this or any other line.)

The distinction between linear and nonlinear data is important for correlation and prediction. We will concentrate on the linear case since it is most commonly used and easiest to understand.

## Prediction

In this section we will discuss the prediction of one measurement from another. Before getting into the prediction problem, recall that there may be a measurement problem that causes our prediction work to be of questionable value. In chapter 2 we discussed the problem of predicting graduate school success. Success is usually measured by some combination of the 4 variables — graduate GPA, comprehensive examination, faculty ratings, and time to degree attainment. If we find a procedure for predicting graduate GPA (one indicator of success), we cannot necessarily conclude that we can predict success in graduate school.

### Explaining Variation

The prediction problem involves us with trying to explain the variation observed in a measurement. Consider, for example, the prediction of a person's weight when everyone in the population of interest is the same weight, that is, there is no variation in weights. The prediction of a person's weight is an easy matter. All one has to do is measure the weight of one person in the population and then predict, without error, the weight of everyone else in the population. So we see that without variation prediction is an easy task.

The previous example is, of course, unrealistic. Not everyone in a population is the same weight. Generally, the numbers in a population are not the same but vary from element to element within the population. If we could explain this variation using, let's say, another measurement, maybe we could predict an unknown quantity with some degree of accuracy.

For example, suppose we tried to predict an adult's weight. The weight of adults varies greatly. It is not likely that we would be able to predict someone's weight with any degree of accuracy. But observe that people who are the same height have weights that vary less that the weights of adults of all heights; that is, adults who are 6 feet tall have weights that vary less than the weights of all adults. Therefore, if we were to be told that a person is 6 feet tall, we should be better able to predict that person's weight.

This is the idea behind prediction. If you want to predict a measurement, find a related measurement that will explain some of the variation that exists in the measurement of interest. We use the known value of the related measurement to predict the unknown measurement.

In the example concerning electrical demand we observe that the variation in daily demand over a year's time would be great. Prediction of demand would be difficult indeed. However, for weekdays in July electrical demand should not vary too greatly for a fixed temperature. Knowing a weekday summer temperature can go a long way toward helping us predict demand.

We will look at this idea in cases in which the measurements of interest are linearly related. The variables of interest are called independent and dependent measurements.

> The *dependent measurement* is the measurement we would like to predict.
>
> The *independent measurement* is the measurement whose *known* value we will use in predicting the level of the corresponding dependent measurement.

In the utility company example temperature is the independent measurement, or the known value, and electrical demand is the dependent variable we would like to predict.

### Fitting a Curve to Data

The prediction problem involves fitting a curve to the data represented by the dots in the scatter diagram. We will consider fitting a line, one type of curve, to the data. Figure 2 shows the graph of the line that best fits the temperature and demand data. The criterion for judging which of, let's say, two lines better fits a set of data pairs is called the *least squares criterion.* One line fits better if the sum of the squared (vertical) distances from the points to the line is smaller than the sum of squared (vertical) distances of the points to the other line. We will discuss this idea later in this chapter.

The lines that might best fit the data sets in figure 3 are also sketched. Nonlinear or curvilinear data, as shown in figure 4, are likely to be better fitted by a curve than by a straight line; both are indicated on that scatter diagram.

Think of fitting a line (or curve) to data as mathematical modeling. We seek a model (line or curve) that best describes the relationship between the independent and dependent measurements. Recall our attempt at modeling the sampling process so we could measure the uncertainty of statistical inference. As then, we must now consider the fit of the data to the model. A linear model does not fit the data graphed in figure 4 as well as a linear model fits the data graphed in figure 2. Sometimes a curvilinear model fits a data set better than a linear model. Sometimes data cannot be accurately modeled at all, see figures 3(g), 3(h), 3(i), and 3(j), for example.

A line (curve) is a graph of points that represents differing values of the two measurements of interest. For example, the line through the data in figure 2 is a graph of temperature and the corresponding electrical output. On this line a temperature of 21.1° is graphed to a demand of 1,041.8, and 26.7° is graphed to 1,174.8. Electrical demand is predicted from temperature by using the relationship between these measurements, which is described by this line; that is, we would predict a demand of 1,041.8 megawatts when the temperature is 21.1°, and a demand of 1,174.8 megawatts for a 26.7° temperature.

A nonlinear curve may also be used in prediction. Consider, for example, the age-weight data shown in figure 5 and the age-height data shown in figure 6.

A pediatrician interprets height and weight data by using graphs such as those shown in figure 5 and figure 6. These graphs give the best-fitting curve for age-weight and age-height data, respectively. A pediatrician will then have some idea of what height or weight to expect for children of different ages. Parents can thus be told if the height or weight of their children is average.

For example, a pediatrician might indicate that a girl is tall and overweight for her age. This is the pediatrician's interpretation of the fact that the girl's height is far above the graph of average heights (figure 6) as well as far above the weight curve given in figure 5. You'll be asked to discuss this interpretation of a girl's height (35.5 inches at age 2) and weight (33 lb at age 2) in the discussion section of this chapter. Observe that according to figure 5 and figure 6, a two-year-old should be about 34 inches tall and weigh about 27 lb.

After a curve is fitted to the data it is used to predict one measurement from the other. An indication of how good such a prediction procedure is would intuitively be the closeness of the data to the curve, that is, how well the data fit the model. For example, for the data represented by the scatter diagrams in figure 3, the prediction procedure would be good for the data in diagrams *a* and *1* since in those cases the data fits a line quite well. Diagrams *e* through *j* of figure 3 indicate a poor prediction of one measurement from the other.

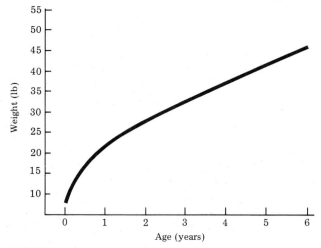

**FIGURE 5    Age-Weight Values**

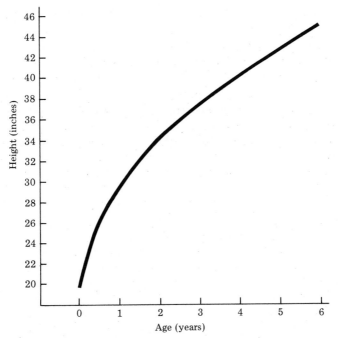

**FIGURE 6**    **Age-Height Values**

An index of linear fit and, hence, an index of how good linear prediction will be is discussed in the next section.

## Correlation Coefficient

The amount, or strength, of a linear relationship, or association, between two measurements is measured by the **correlation coefficient.** A **positive,** or **direct,** linear association is the tendency of larger (above average) values of one measurement to be paired with the larger (above average) values of the second measurement and the smaller values with the smaller. The temperature and electrical demand data of table 2 has such a tendency. A scan of these data indicates that the larger temperatures tend to be paired with the greater electrical demand. A **negative,** or **inverse,** linear association is the tendency of the larger values (above average) of one measurement to be paired with the smaller values (below average) of the other measurement and smaller values to larger values. An example of data with a negative relationship would be the age of a car and its value. Excluding models with antique value, the older cars would be paired with smaller values.

A frequently used measure, or index, of the strength of a linear relationship is the ***Pearson product-moment correlation coefficient.*** This correlation coefficient is a measure of the tendency of large (above average) values of one measure to be paired with above average values of another measurement or of the tendency of large values of one measurement to be paired with small (below average) values of the other measurement. (This statistic is often denoted by the letter *r* for "reversion." The letter was originally used by Galton to represent the slope of the regression or prediction line. Other correlation coefficients are also used.) The correlation coefficient is always a number between a negative 1 (−1) and a positive 1 (+1). Values near −1 indicate high negative linear correlation. A correlation coefficient near 0 indicates no linear relationship, and values near +1 indicate high positive linear correlation. The data of figure 3 have corresponding correlation coefficients indicated on each scatter diagram.

Note that there could be a high curvilinear association, although the measure of linear association might be near 0. (We will consider this phenomenon in Exercise 2.)

### Interpretation

We must be careful when we interpret the correlation coefficient. A correlation coefficient of 0.6 does not represent twice as much association, or relationship, as a correlation coefficient of 0.3. We can say, however, that:

1.  The correlation coefficient is positive or negative depending on whether the linear relationship is positive or negative, respectively.

2.  A coefficient of +1 or −1 means perfect positive or perfect negative linear relationship, respectively.

3.  If no linear relationship exists, the correlation coefficient is 0. We are not saying, however, that a correlation coefficient of 0 means that no relationship exists between the measurements.

4.  The closer to +1 or −1, the stronger the linear relationship between the variables.

A correlation coefficient is meaningful only with respect to the range of the two measurements involved. If the range of the measurements is changed, the correlation coefficient necessarily changes. For example, a correlation coefficient of .89 between voter turnout percentages and voter registration percentages was reported for 104 cities.[4] However, the correlation coefficient for the nine data pairs with voter registration between 90% and 100% was only .38.

When interpreting the correlation coefficient, be careful of using the words "high," "medium," and "low" for describing the existing relationship. A correlation coefficient of 0.3 could be high or low depending on the measurements under consideration. A coefficient of 0.3 could be quite high, for example, if we were looking at weight versus intelligence. However, a correlation coefficient of 0.3 might be quite low when, let's say, two physical characteristics such as height and weight are being measured. This follows since a physical measurement is not likely to be related to a measurement of a mental trait. Hence, a correlation coefficient of 0.3 could be high for such measurement pairs when compared with the correlation coefficient usually obtained when looking at data that match a physical and a mental characteristic.

On the other hand, two physical measurements such as height and weight may often have a large correlation coefficient. If two physical measurements usually have a strong linear relationship, a correlation coefficient for 0.3 for such measurements could be low. In any case we must know the magnitude of correlation coefficients that have been found for similar measurement pairs before making a judgment that a particular value is high or low.

The correlation coefficient is also an index of the accuracy of linear prediction. This statistic indicates how well actual values of the dependent measurement cluster about the prediction, or regression, line. Since we predict values of this measurement by points on the line, the correlation coefficient will then tell us how close the predicted value tends to be to the actual measurement; that is, the closer the correlation coefficient to $+1$ ($-1$ for a negative relationship), the better the prediction of one measurement from the other.

The square of the correlation coefficient also indicates the amount of variation in the dependent variable that is explained by the independent variable. Recall that we want to be able to explain the variation observed in the dependent measurement by knowing the value of the corresponding independent measurement. The square of the correlation coefficient gives us this information. For example, the correlation coefficient for the temperature and output data is about .86. The square of .86 is .74; that is, 74% of the variation in electrical demand can be explained by knowing the temperature.

## Testing Independence

Situations arise in which the observed pairs may be thought of as a sample from a population. For example, a sample of graduate students may be selected. Information on the relationship between undergraduate GPA and graduate GPA might shed some light on the relationship between these variables for the student population from which the sample was drawn. For the electrical output and temperature data the utility company may want to infer from the July 1971 data to other summer months during other years. Caution would have to be urged here, however, since there is no information on month to month variation during a particular summer, nor are there data on the year to year changes that might affect demand. Background influences on statistical inferences, such as to which population inference may properly be made, must always be considered.

Let us look now at one important test of hypothesis — the test for no association. We will look at both the continuous and categorical data cases.

### Test for No Association: Continuous Data

The Pearson product-moment correlation coefficient can be thought of as a statistic arising from a sample of data pairs selected from a population. A *population correlation coefficient* does exist, although we may not know its value.

A natural question to ask is whether the population correlation coefficient is zero. If the population correlation coefficient is zero, this would mean that the two variables are unrelated in a linear manner; that is, the independent measurement tells us nothing at all about the dependent variable as far as a linear relationship is concerned. If we would like to predict the dependent variable when the variables are not related, we might just as well look at the dependent variable by itself since the independent variable tells us nothing about the variation we observe in the dependent variable. For example, suppose the correlation coefficient between IQ and height is zero. (It probably is.) If we want to predict a person's IQ, we will not be aided by knowing the person's height. In contrast, since height and weight are related, that is, the correlation coefficient is greater than zero, knowing a person's weight will aid us in predicting his or her height. The test that a correlation coefficient is zero is important as we need to know if the extra information contained in the independent variable is worth collecting.

null hypothesis: population correlation coefficient = zero

The test of whether a correlation coefficient is zero is rather straightforward. To find the sample statistic we will use in the test, we multiply the square root of one less than the sample size times the sample correlation coefficient:

test statistic = (sample correlation coefficient)
$\times$ square root (sample size − 1)

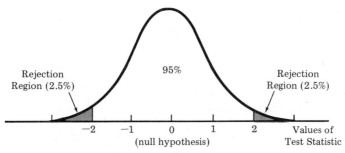

**FIGURE 7**    **Distribution of Test Statistic for Testing for Linear Relationship**

This product has an approximated standard normal sampling distribution if the null hypothesis is true; that is, under the null hypothesis, the test statistic has an approximately normal distribution with mean 0 and standard error 1 as shown in Figure 7.

Recall that a 5% standard normal test would be:

1. Reject the null hypothesis if the standard normal statistic exceeds 2 or is less than $-2$.

2. Fail to reject the null hypothesis if the standardized statistic is between $-2$ and 2.

As an example, suppose data are collected on a sample of students. The data concern the amount of alcohol consumed in a specified period and the amount of marijuana smoked during the same period. The sample correlation coefficient was found to be .565 for 24 students. We then need to know the square root of $(24 - 1)$, or 23, which is 4.8.

We are testing whether the correlation coefficient between alcohol consumption and marijuana use is zero; that is, we are testing whether the amount of alcohol a person consumes tells us anything about the amount of marijuana the person smokes. We reject the null hypothesis of no relationship if the test statistic is less than $-2$ or exceeds 2. Our standard normal test statistic is

$$.565 \times 4.8 = 2.7$$

We therefore reject the null hypothesis. Alcohol consumption can tell us something about marijuana use: People who drink more alcohol are likely to smoke more marijuana.

### Test for No Relationship: Categorical Data

As with the categorical data discussed in Chapter 2, data pairs may also be categorized. Each of the two variables is categorized into one of any number of classes. For example, as we discussed in chapter 7, in a study on smoking and lung cancer people may be categorized by "smoke" or "don't smoke" and "lung cancer" or "no lung cancer" — that is, people are put in one of 4 categories as summarized in table 2.

Recall our discussions of a contingency table in chapter 7. Our sampling plan is as for a point-in-time study. A fixed sample of people are selected and counted in one of the 4 cells of the contingency table: each cell of the table represents presence or absence of each of two dichotomous variables.

We might assume that the data collected are a sample of pairs from a population. We wish to ask if the variables of interest are related. In the smoking and cancer data we might wonder if smoking and lung cancer are related.

Let us look at the details of running a test for independence of two categorical variables. We will look at a sample of students who may or may not smoke marijuana and may or may not drink alcoholic beverages. The data are shown in table 3. Hence, of the 330 students interviewed 193 both drank alcoholic beverages and smoked marijuana.

If drinking and smoking are unrelated, we would expect the percentage of marijuana smokers who drink to equal the percentage of nonsmokers of marijuana who drink. We see from table 3 that in total there are 215 of 330 people who drink. This percentage of drinkers,

$$\frac{215}{330} = 65\%$$

**TABLE 2    Categorization (Contingency Table)**

|  |  | Smoke | |
|---|---|---|---|
|  |  | Yes | No |
| Lung Cancer | Yes |  |  |
|  | No |  |  |

**TABLE 3    Drinking and Marijuana Data: Observed Frequencies**

|  |  | Smoke Marijuana | | |
|---|---|---|---|---|
|  |  | Yes | No | Totals |
| Drink | Yes | 193 | 22 | 215 |
|  | No | 71 | 44 | 115 |
|  | Totals | 264 | 66 | 330 |

**TABLE 4**    Drinking and Marijuana Data: Expected Frequencies

|  |  | Smoke Marijuana | | |
|---|---|---|---|---|
|  |  | Yes | No | Totals |
| Drink | Yes | 65% of 264<br>172 | 65% of 66<br>43 | 215 |
|  | No | 35% of 264<br>92 | 35% of 66<br>23 | 115 |
|  | Totals | 264 | 66 | 330 |

**TABLE 5**    Summary of Calculations of Test Statistic

|  |  | Smoke Marijuana | |
|---|---|---|---|
|  |  | Yes | No |
| Drink | Yes | $\dfrac{(193 - 172)^2}{172} = 2.56$ | $\dfrac{(22 - 43)^2}{43} = 10.26$ |
|  | No | $\dfrac{(71 - 92)^2}{92} = 4.79$ | $\dfrac{(44 - 23)^2}{23} = 19.17$ |

should be the same for both users and nonusers of marijuana. Hence, 65% (or 172) of the total 264 users of marijuana would be expected to also drink. Similarly, 65% of the 66 nonusers of marijuana,

$$65\% \text{ of } 66 = 43$$

would drink. Table 4 summarizes what we would expect to observe if smoking marijuana and drinking are unrelated.

   To derive the statistic to test for no association (independence) we look at the squared deviations between the observed frequencies (given in table 3) and the expected frequencies (given in table 4). Then, the squared deviations are divided by the expected frequencies and added over the 4 categories (cells).

$$\text{Test statistic: sum of } \frac{(\text{observed} - \text{expected})^2}{\text{expected}}$$

Table 5 summarizes these calculations. The sum of the standardized squared deviations is

$$2.56 + 10.26 + 4.79 + 19.17 = 36.78$$

This statistic has a chi-square (1) distribution. Recall from our discussions in chapter 4 that the chi-square distribution applies to continuous data. Although the original data is categorical, the distribution of the test statistic can be closely approximated by the chi-square distribution. Referring to Appendix B, table 2 and recalling our discussion of a chi-square distribution in chapter 4, we set up a 5% test as:

1. Reject the null hypothesis if the statistic exceeds 3.8.

2. Fail to reject the null hypothesis if the statistic is less than or equal to 3.8.

Since our statistic is 36.78, we reject the hypothesis of independence; that is, drinking and marijuana are related — a person who drinks is more likely to also smoke marijuana.

## Cause-and-Effect versus Linear Relationships

The source of much misinterpretation of statistics in the area of correlation and prediction is the inferring of a cause-and-effect relationship between characteristics from the existence of a high linear relationship. A correlation coefficient near $-1$ or $+1$ indicates a high linear relationship. This linear association is as we described it before: For a positive relationship large (above average) values of one measurement tend to be paired with large (above average) values of the other measurement. It is wrong, however, to infer a cause-and-effect relationship on the basis of a high linear relationship alone, even if a significant correlation exists.

For example, a strong linear relationship exists between the height and the achievement of grade school students (K–6). It would be ludicrous to infer that greater height causes higher achievement (i.e., make me taller so I can be smarter). This mathematical relationship arises from the fact that the taller students in this group tend to be older and, hence, have higher achievement.

The noted statistician, Jerzy Neyman, tells of a person who found a high positive linear relationship between the birth rates in counties and the density of storks in the counties. Should we therefore conclude that storks bring babies after all? (This is a spurious correlation as the number of births and the number of storks are reported as percentages with respect to the number of women in the counties. There is no relationship between the number of births and the number of storks except that which arises spuriously by dividing both of these numbers by the same value — the number of women in the county.)[5]

Recall our discussion in chapter 7 about the controversy over claims that smoking causes lung cancer since there is a high (positive) relationship between amount of smoking and incidence of lung cancer. This fact by itself does not prove a cause-and-effect relationship. (See Discussion 5.)

Be careful when trying to interpret measurements that display a high relationship, linear or nonlinear. There may be explanations for this relationship other than the existence of a cause-and-effect relationship.

## Calculating a Correlation Coefficient and Determining the Prediction Line (Optional)

Consider the following 5 data pairs:

(3.8, 18), (3.0, 13), (1.9, 26), (1.9, 28), and (1.6, 28)

The first value in each pair represents a student's freshman GPA, and the second measurement in each pair is the American College Testing (ACT) score, a standardized entrance exam for undergraduate school, which is similar to the GRE for graduate school entrance.

To calculate the Pearson product-moment correlation coefficient we look at the deviation of each measurement from the mean of that measurement — that is, we first consider the mean of the 5 GPA and the 5 ACT scores. The mean GPA of the 5 students is 2.44, and the mean ACT score is 22.6.

$$\frac{3.8 + 3.0 + 1.9 + 1.9 + 1.6}{5} = \frac{12.2}{5} = 2.44$$

$$\frac{18 + 13 + 26 + 28 + 28}{5} = \frac{113}{5} = 22.6$$

The deviations from the means are given in table 6.

TABLE 6    Calculation of the Correlation Coefficient

| I | II | III | IV | V |
|---|---|---|---|---|
| | Deviation of GPA from | | Deviation of ACT from | Product of Columns II and |
| Freshman GPA | mean GPA | ACT Scores | mean ACT | IV |
| 3.8 | 1.36 | 18 | −4.6 | −6.256 |
| 3.0 | .56 | 13 | −9.6 | −5.376 |
| 1.9 | −.54 | 26 | 3.4 | −1.836 |
| 1.9 | −.54 | 28 | 5.4 | −2.916 |
| 1.6 | −.84 | 28 | 5.4 | −4.536 |
| | | | Total: | −20.920 |

The mean of the products given in column V is $-4.184$ $(-20.92/5 = -4.184)$. Observe that the products in column V reflect the tendency of large GPAs (above average GPAs) to be paired with low ACT scores (below average) and low GPAs to be paired with low ACTs.

We standardize this average by dividing it by the standard deviation of the GPAs times the standard deviation of the ACT scores. These standard deviations are .83 and 6.05, respectively. Recall our discussion in chapter 3 that the mean of the squares of the numbers given in column II and column IV of table 6 gives the variances of the GPA and ACT scores, respectively. The standard deviations are the square roots of these variances. The correlation coefficient is then

$$\frac{-4.184}{(.83 \times 6.05)} = -.83$$

(see Discussion 7).

Using the summary statistics given in table 6, we can also determine the least-squares-fitted line — that is, determine the best linear fit of the data using the least squares criterion. Any line can be completely specified by knowing its slope and intercept. The intercept indicates where the line crosses the vertical axis on a graph, and the slope indicates how much the line is tilted. The slope of this line is given by

$$\text{slope} = \text{correlation coefficient} \times \frac{\text{standard deviation of dependent variable}}{\text{standard deviation of independent variable}}$$

In our example

$$\text{slope} = -.83 \times \frac{.83}{6.05} = -.1$$

The intercept, the point where the line intersects the vertical scale, is given by

$$\text{intercept} = \text{mean of dependent variable} - (\text{slope} \times \text{mean of independent variable})$$

In our example

$$
\begin{aligned}
\text{intercept} &= 2.44 - (-.1 \times 22.6) \\
&= 2.44 + 2.26 \\
&= 4.7
\end{aligned}
$$

The equation for the line is given by

dependent variable = intercept + (slope × independent variable)

If the independent variable took the values 15, 20, and 25, the dependent variable would be as shown in table 7. If the three points given in table 7, (15, 3.2), (20, 2.7), and (25, 2.2), are indicated on a scatter diagram and then connected by a line, the resulting line is the least-squares-fitted line shown in figure 8. (Any two points will fix or determine a line. It's good practice to graph three points so that an error is less likely made — all three points should fall on a line.) The original 5 data pairs are indicated by circles in the scatter diagram in figure 8.

**TABLE 7    Calculation of Points on the Regression Line**

| Independent Variable | Dependent Variable |
|---|---|
| 15 | intercept + (slope × 15) = 4.7 + (−.1 × 15) <br> = 4.7 − 1.5 = 3.2 |
| 20 | 4.7 + (−.1 × 20) = 4.7 − 2 = 2.7 |
| 25 | 4.7 + (−.1 × 25) = 4.7 − 2.5 = 2.2 |

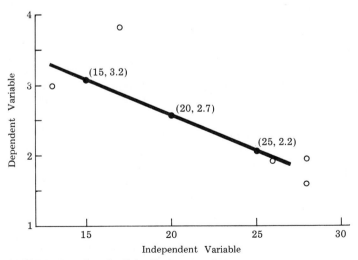

**FIGURE 8    Graph of the Regression Line**

## Summary

The relationship between two different characteristics is often of interest to us as we may be able to predict one characteristic from the other. Data pairs can be represented graphically by means of a scatter diagram. From such a graph we may describe the relationship between the measurements as linear (elliptical) or curvilinear.

The correlation coefficient is a measure of the strength of a linear relationship. Interpretation of this statistic is not at all clear-cut. We could, however, say that:

1. The correlation coefficient is positive or negative depending on whether the relationship is direct or inverse, respectively.

2. A correlation coefficient of $+1$ indicates a perfect positive linear relationship; minus one indicates perfect negative linear relationship.

3. When no linear relationship exists, the correlation coefficient is 0.

4. The closer the correlation coefficient is to $+1$ (or $-1$), the stronger the linear relationship between the variables.

We hedged on using adjectives like "low," "medium," and "high" to describe a relationship. A correlation coefficient of a certain level may not always indicate, let's say, a high relationship. What is a high coefficient for certain pairs of measurements may not be a high coefficient for other measurement pairs.

After fitting a line (curve) to data we can predict one measurement from the other measurement. This is a commonly used method for predicting success in school or at a job. We looked in some detail at the prediction of electrical demand using temperature. The correlation coefficient is an index of the accuracy of linear prediction. The closer the correlation coefficient is to $+1$ or $-1$, the better the linear prediction.

The test of independence was discussed for both continuous and dichotomous pairs of data. Assuming we had sampled a population of pairs of continuous observations, we looked at a normal test that the population correlation coefficient is zero. For data pairs in which each of the two measurements of interest was dichotomous we discussed a chi-square test that the traits being measured are not related, that is, the traits are independent.

Proving that a relationship does indeed exist was not sufficient to show a cause-and-effect relationship between the characteristics being studied. As we discussed in chapter 7, a causative relationship is quite difficult to prove. Rejection of a hypothesis of independence is but one piece of information that could be used to prove causality.

# No Comment

■ *The age-weight and age-height data discussed in the "Fitting a Curve to Data" section of this chapter and in Discussion 1 may have come from a 1945 article of the Journal of Pediatrics.*
*(Charts copyrighted 1943 by State University of Iowa)*

■ *Football player explaining why he didn't play major-college football his freshman year: "I couldn't predict a 1.6 grade average [so he wasn't eligible]."*

# Exercises

1. Would there be positive, negative, or no linear relationship between the following pairs of measurements?
   a. The number of hours a golfer practices and his or her score
   b. A person's shoe size and his or her score on an art appreciation test
   c. Amount of tread on a tire and the number of miles it has been driven
   d. Car weight and gas mileage

2. Consider the 5 data pairs (0,3), (1,2), (2,0), (3,2), and (4,3). The correlation coefficient is 0. (You may want to verify this using the techniques in the last section of this chapter.) Graph these data and describe the relationship between the two measurements represented in the data pairs. Recall what the correlation coefficient measures.

3. The number of "defective" items are recorded for each of two days in table 8. Does the number of defectives change from day to day, that is, is the quality of the items independent of day?

TABLE 8    Number of Defective Items

|  |  | Day | |
|---|---|---|---|
|  |  | 1 | 2 |
| Defective | Yes | 108 | 78 |
|  | No | 12 | 22 |

**TABLE 9**    **Number of Accidents**

|           | 10:00–11:00 AM | 2:00–3:00 PM |
|-----------|----------------|--------------|
| Monday    | 11             | 8            |
| Wednesday | 9              | 13           |

**TABLE 10**    **Number of Applications**

|        |   | School |          |
|--------|---|--------|----------|
|        |   | Law    | Medicine |
| School | 1 | 30     | 58       |
|        | 2 | 36     | 52       |

**TABLE 11**    **Advertising — Sales Data**

| Sales       | 13 | 9 | 8 | 13 | 7 | 10 | 11 | 9 |
|-------------|----|---|---|----|---|----|----|---|
| Advertising | 4  | 1 | 4 | 8  | 2 | 6  | 7  | 8 |

4.  Table 9 illustrates the records of a safety engineer regarding the number of accidents that occur in a particular plant on Monday and Wednesday during the periods of 10:00 AM to 11:00 AM and 2:00 PM to 3:00 PM. Does the number of accidents change from day to day?

5.  Records at two universities were sampled. Information on the number of applicants to law school and medical school were recorded as shown in table 10. Is there an association between the university and the number of applicants for these schools?

6.  Given the data from the last section of this chapter, is there a significant correlation between freshman GPA and ACT scores?

7.  (Optional) Data were collected on the sales of a product for different levels of advertising. The data are recorded in terms of thousands of dollars in table 11. Describe the relationship between the two variables. What is the correlation coefficient? Is there a significant correlation? What is the least-squares-fitted line? Graph this line.

# For Discussion

1.   Discuss the height-age and weight-age data mentioned in the section on prediction in this chapter. Specifically, how should we interpret a girl's weight of 33 lb and height of 35.5 inches at age 2 years. (Recall figure 5 and figure 6.) Any background influences? Suppose these data came from a 1945 study.

   As an aid to your discussions, consider the graphs in figure 9 and figure 10. The broken bands in these graphs are 68% *confidence bands* for the two data sets. The solid lines represent 95% confidence bands. The confidence band should be interpreted as a confidence interval (see chapter 6). With this band a confidence interval is represented for each value on the horizontal score scale. The dot on each graph represents 2 years, 33 lb and 2 years, 35.5 inches, respectively.

2.   Discuss the use of GRE scores for predicting graduate school success. How well might you expect to predict success for students with low GRE scores (scores less than most students in graduate school)?

3.   It is not unusual to estimate flood damage for different stages of a river by fitting a curve to the flood damage (in dollars) in

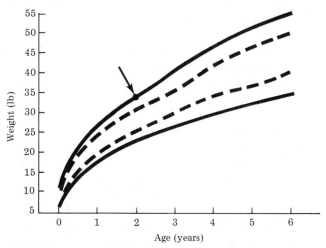

**FIGURE 9    Confidence Bands for Age-Weight Data**

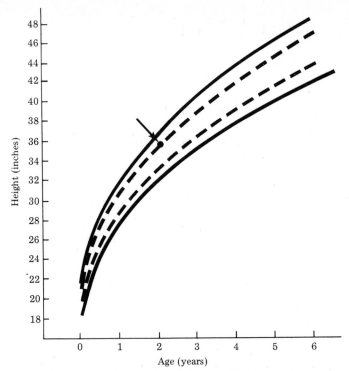

**FIGURE 10    Confidence Bands for Age-Height Data**

relationship to river levels for floods that have occurred. For example, three floods have occurred in a particular area as indicated in figure 11.

Discuss the accuracy of predicting flood damage for different flood stages. Since the floods cited in figure 11 occur in different years, does a "dollar" mean the same thing at all three times? (We will return to this problem in chapter 10.)

4.   The correlation coefficient between ACT scores and freshman GPAs was found to be $-.83$ for the data given in the last section of this chapter. Discuss what this descriptive statistic means. Inferentially, how credible is this value?

5.   Reconsider the controversy over whether smoking causes lung cancer in light of the relationship mentioned at the end of this chapter between the amount of smoking and incidence of lung cancer. (Recall the section on causation in chapter 7 as well as Discussion 6 and Exercise 1 in chapter 7.)

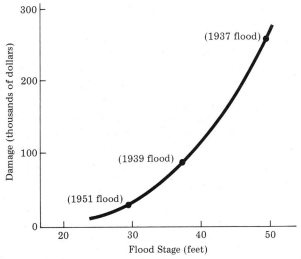

**FIGURE 11**   Flood Level-Damage Graph

6.  In practice more than one independent measurement may be used to predict the value of an independent measurement. For example, the voting results of many (predictor) precincts may be used to predict the results of an election on election night before all the ballots are counted. (Recall our discussion in chapter 6 of this testing problem.) How would precincts be selected in this instance? Compare this to the selection of precincts in the example presented in the section on testing a statistical hypothesis in chapter 6.[6]

7.  Earlier in this chapter we ran tests concerning the relationship between drinking alcoholic beverages and smoking marijuana. In one instance we considered drinking and smoking as continuous variables, and in the other case we looked at dichotomous, ("yes"–"no") responses. Discuss the background influences on these two sets of data pairs. Which set is likely to be more reliable?

# Further Readings

The articles in Tanur's *Statistics: A Guide to the Unknown* that discuss prediction and correlation are recommended reading. In particular, consider the article by E. R. Tufte, "Registration and Voting" and the article by R. F. Link, "Election Night on Television." Details of publication are given in the chapter Notes.

# Notes

1. A. L. Dudycha and L. W. Dudycha, "Behavioral Statistics: An Historical Perspective," *Statistical Issues: A Reader for the Behavior Sciences,* ed., R. E. Kirk (Monterey, Calif.: Brooks/Cole, 1972) p. 17.
2. F. Galton, *Memories of My Life* (London: Methuen, 1908), p. 300.
3. *Ibid.*
4. E. R. Tufte, "Registration and Voting," *Statistics: A Guide to the Unknown,* ed., J. M. Tanur (San Francisco: Holden-Day, 1972), pp. 153–161.
5. J. Neyman, "Lectures and Conferences on Mathematical Statistics and Probability," Graduate School, U.S. Department of Agriculture, part 3, (1953) pp. 43–46.
6. R. F. Link, "Election Night on Television," *Statistics: A Guide to the Unknown,* ed., J. M. Tanur, pp. 137–145.

# Index Numbers

10

INDEX NUMBERS ARE WIDELY USED in business and economics. For example, the use of the Consumer Price Index has become so extensive that your wages may now be tied to this statistic and the benefits you will receive at retirement through the Social Security system will certainly be tied to this index. In this chapter we will investigate these types of statistics so you can better understand the statistical information contained in them.

We will consider how index numbers are calculated as well as how to use an index number such as the Consumer Price Index to change dollar amounts to fixed, or constant, dollars. The buying power of the dollar is continually changing. We must therefore convert dollar quantities to a fixed dollar such as 1967 dollars before we can compare dollar quantities from different time periods. By converting to 1967 dollars we could, for example, properly evaluate a change in pay. In this case we are comparing dollar amounts for two different years. The buying power of the dollar decreases from one year to the next. If a pay change does not keep pace with changes in the cost of goods and services we buy, we may be getting a decrease in pay rather than an increase in pay.

## Definition, History, and Construction of Index Numbers

> Numbers that reduce data to values relative to the same quantity or base are called *index numbers*.

The phrase "index numbers" was first used in 1869 by the *London Economist* to refer to the sum of the price relatives of 22 commodities. A *price relative* is the change in cost of a product with respect to some base time. The value is calculated with the formula

$$\frac{\text{current price}}{\text{base price}} \times 100$$

The factor of 100 is used to change the ratio to a percentage. For example, a product whose cost increased from 6¢ to 8¢ per unit would have a price relative of

$$\frac{8}{6} \times 100 = 133\%$$

Index numbers are used quite often to compare prices, production, employment, or population changes over a period of time. Index numbers may be thought of as percentages because in their simplest form this is what they are. Today many different index numbers exist including the Consumer Price Index, the Wholesale Price Index, the FBI's Uniform Crime Index, the Gross National Product, and employment/unemployment figures.

The use of index numbers to measure price changes began with the work of Count Gian Rinaldo Carli (1720-1795). With the opening of the New World to trade in the late eighteenth century came an influx of gold and silver, which was the apparent reason that prices in Europe spiraled upward. It has been estimated that in the 1700s the quantity of metal money in Europe quadrupled. Most economists of this period believed that the increase in prices was due to the increase in the supply of gold and silver. Carli disagreed, contending that the rise in prices was a consequence of the debasement (devaluation) of coinage rather than of the increase in bullion supply. He offered empirical evidence in support of his views by comparing the purchasing power of a unit of bullion in Italy at the time of the discovery of America with the contemporary unit (about 1750). In another treatise "he described the monetary systems of Italy and showed that the bullion content of coinage had declined because the rulers of Italy debased their currencies to finance lavish spending on palaces, churches, paintings and monuments. The diminishing supply of gold in Italy was attributed by Carli to an unfavorable balance of trade."[1] An important by-product of Carli's work was the earliest recorded use of index numbers.

The problem of constructing a price index begins with considering which items are to be priced. Carli priced only three items — wheat, wine, and olive oil.

A second consideration is the selection of a base period, which is a period of relative normality. We are speaking of normality with respect to prices if a price index is to be constructed. Price change data will then be transformed to numbers relative to this base period. Carli selected the time of the discovery of America as his base period. Carli's choice of this time as a base period was reasonable since the concern of economists at that time was how the influx of gold and silver from the New World affected prices in Italy. An index number is usually set at 100 for the base period so that later price changes may be interpreted as percentage changes relative to the base period.

A third consideration involved in the construction of an index is a technique for weighting items. For example, a price change in meat would be weighted more if more meat is consumed than another item to be priced, for example, cereal.

Suppose that in a certain area three half-gallons of milk are purchased for each dozen eggs. Also, suppose that the price of milk changed from 76¢ to 82¢ a half-gallon and eggs increased from 64¢ to 72¢ a dozen during the same period of time; that is, milk increased

$$\frac{82 - 76}{76} \times 100 = 8\%$$

and eggs increased

$$\frac{72 - 64}{64} \times 100 = 12.5\%$$

A value representing the earlier prices of these items and also reflecting the difference in the quantities of the items purchased would be

$$(3 \times \$.76) + (1 \times \$.64) = \$2.92$$

Think of this as the price of three half-gallons of milk and one dozen eggs during the base period. The weighting values, three and one, reflect the fact that people in the area of interest buy three half-gallons of milk for each dozen eggs. A corresponding value for the more recent time period would be

$$(3 \times \$.82) + (1 \times \$.72) = \$3.18$$

A weighted price index, which reflects the change in the price of milk and eggs over the period of interest, would be

$$\frac{3.18 - 2.92}{2.92} \times 100 = 9\%$$

The price of these items increased 9% over the time period of interest.

We will now take a look at a particular price index. The Consumer Price Index or the "billion-dollar index" as it is often called, will be discussed in detail in the next section.


## The Consumer Price Index

The Consumer Price Index (CPI) is presently a measurement of price changes of goods and services purchased by a particular segment of our population. Its official name is the *Consumer Price Index for*

*Urban Wage Earners and Clerical Workers*. Its name prior to January 1964 was the *Index of Change in Prices of Goods and Services Purchased by City Wage-Earner and Clerical-Worker Families to Maintain Their Level of Living*. It is from this title that the CPI became known as the "cost-of-living" index. But the CPI is not a cost-of-living index. It measures only price changes of goods and services bought by wage earners and clerical workers as a broad group but not necessarily the price changes of items bought by any one family or small group of families.

Our cost of living may be affected by factors not measured by the CPI such as changes in family composition. Also not measured by the CPI are the changes in family buying habits. For example, due to a rise in the price of beef many families will eat more pork, fish, and cheese, rather than pay the higher beef prices. Therefore, when the CPI shows an increase in beef prices, it may not truly reflect the cost of living for families who had adjusted their buying habits.

## History and Influence

The CPI was initiated during World War I when a measure of living costs, particularly in shipbuilding centers, became essential in wage negotiations. The Bureau of Labor Statistics began publishing the index in 1919. The use of the CPI declined during the post–World War I and depression periods but was revived again during World War II. It did not receive widespread acceptance until 1948, when a contract between the United Automobile, Aircraft, and Agricultural Implement Workers of America and General Motors Corporation contained the first major use of the so-called *escalator clause*. Such a clause makes a pay raise automatic with an increase in the CPI. In 1975 nearly 5.2 million workers were under contracts containing an escalator clause.

Social Security rates are now tied to the CPI. As the CPI increases, so do the nearly 29 million Social Security checks sent monthly to the retired citizens, widows with children, and disabled of our country. The Social Security system is supposed to be self-supporting, that is, the taxes taken from workers' paychecks should cover payments to Social Security recipients. With rates set to the CPI, when inflation (as measured by the CPI) increases faster than wages, the Social Security system could cease being self-supporting.

The benefits paid to about 2 million retired military and federal workers, 600,000 postal workers, and 17 million food stamp recipients are also escalated with increases in the CPI.

In addition the CPI has also become widely used in government policy making, since it is an inflation indicator. So widespread is the use of the CPI in labor management negotiations and government policy making that a 1951 congressional committee on the CPI called it a "billion-dollar index." In fact, it is estimated that a 1% increase in the CPI triggers at least a $1 billion increase in incomes under wage escalator provisions.

An index with such a strong influence on our lives must be understood. We will first learn what the CPI is designed to measure.

### What the CPI Measures

Since the full name for the CPI is the *CPI for Urban Wage Earners and Clerical Workers*, our first concern is with exactly who are considered to be urban wage earners and clerical workers:

> The definition of wage earners and clerical workers is based on the occupational classifications used by the Bureau of Census for the 1960 Census of Population and listed in the Alphabetical Index of Occupations and Industries. The group includes *craftsmen, foremen*, and *kindred workers*, such as carpenters, bookbinders, etc.; *operatives* and *kindred workers*, such as apprentices in the building trades, deliverymen, furnacemen, smelters and pourers, etc.; *clerical* and *kindred workers*; *service workers, except private household*, such as waitresses, practical nurses, etc.; *sales workers*; and *laborers* except *farm and mine*. It excludes *professional, technical*, and *kindred workers*, such as engineers and teachers; *farmers and farm managers; managers, officials* and *proprietors, except farm; private household workers*; and *farm laborers* and foremen. A consumer unit included in the 1960–61 Survey of Consumer Expenditures was classified in the index group if more than half the combined income of all family members was obtained in a wage earner and clerical worker occupation and at least one family member was a full-time earner (that is, worked 37 weeks or more during the survey year).[2]

Wage earners and clerical workers make up about 40% of the labor force, 55% of the urban population, and 45% of the total United States population. (Recall from our discussions in chapter 2 that the labor force is made up of employed and unemployed people. Refer to this chapter to refresh your memory on who are counted as employed and who are counted as unemployed.) About two thirds of the individuals classified in this category are males. The mean size of the families included is about 3.7 persons, and the mean family income is about $6,230 after taxes. The mean income of single persons included in this group is about $3,560 after taxes.

Since the CPI does not reflect the price changes of goods and services bought by teachers who are bargaining collectively with school districts, teachers may soon be included in the calculation of the CPI. If teachers' contracts should begin to include escalator clauses, teachers should be included in CPI calculation.

A major change in the CPI took effect 27 February 1978. A second index representing the spending experiences of 80% of the total United States population including some urban wage earners and clerical workers is now available. This new index reflects changes in prices of goods and services bought by all urban families, the elderly, the poor, the self-employed, and the unemployed. Both the new (larger) index and the CPI are published simultaneously.

Another consideration of interpretation is the goods and services that are priced in the CPI. The goods priced for the CPI come from a market basket of items. All items bought by an eligible family may be part of the market basket. In reality only priceable items make up the market basket. Priceable items are items that can be described clearly, are bought and sold with some degree of regularity, and are bought in sufficient quantities to have a measurable price. Table 1 gives an indication of some of the items priced for the CPI. Services such as public transportation, health, and recreation are included.

TABLE 1    CPI; Relative Importance of Major Groups, Special Groups, and Major Subgroups—December 1971 and December 1970

| | Percent of All Items | |
|---|---|---|
| Components | December 1971 | December 1970 |
| All Items | 100.00 | 100.00 |
| *Major Groups* | | |
| Food | 22.19 | 21.99 |
| Housing | 33.84 | 33.80 |
| Apparel and Upkeep | 10.45 | 10.57 |
| Transportation | 13.27 | 13.53 |
| Health and recreation | 19.87 | 19.73 |
| Medical care | 6.46 | 6.37 |
| Personal care | 2.58 | 2.60 |
| Reading and recreation | 5.71 | 5.66 |
| Other goods and services | 5.12 | 5.10 |
| *Special Groups* | | |
| All items less shelter | 78.28 | 78.18 |
| All items less food | 77.81 | 78.01 |
| All items less medical care | 93.54 | 93.63 |
| Commodities | 62.59 | 62.85 |
| Nondurables | 45.82 | 45.80 |
| Durables | 16.77 | 17.05 |
| Services | 37.41 | 37.15 |

TABLE 1 — *Cont'd*

| | Percent of All Items | |
|---|---|---|
| Components | December 1971 | December 1970 |
| Commodities less food | 40.40 | 40.86 |
| Nondurables less food | 23.63 | 23.81 |
| Household durables | 4.83 | 4.90 |
| House furnishings | 4.20 | 4.25 |
| Services less rent | 32.36 | 32.12 |
| Household services less rent | 15.36 | 15.34 |
| Transportation services | 5.57 | 5.47 |
| Medical care services | 5.55 | 5.45 |
| Other services | 5.88 | 5.86 |
| *Major Subgroups* | | |
| Food | 22.19 | 21.99 |
| Food at home | 17.23 | 17.08 |
| Cereals and bakery products | 2.28 | 2.31 |
| Meats, poultry, and fish | 5.04 | 5.57 |
| Dairy products | 2.72 | 2.76 |
| Fruits and Vegetables | 3.03 | 2.78 |
| Other food at home | 3.56 | 3.66 |
| Food away from home | 4.96 | 4.91 |
| Housing | 33.84 | 33.80 |
| Shelter | 21.72 | 21.82 |
| Rent | 5.05 | 5.03 |
| Hotels and Motels | .42 | .42 |
| Home-ownership | 16.25 | 16.37 |
| Fuel and utilities | 4.71 | 4.59 |
| Fuel oil and coal | .67 | .67 |
| Gas and electricity | 2.43 | 2.35 |
| Telephone, water, and sewer | 1.61 | 1.57 |
| Household furnishings and | | |

| | Percent of All Items | |
|---|---|---|
| **TABLE 1** — *Cont'd.* *Components* | *December 1970* | *December 1971* |
| operation | 7.41 | 7.39 |
| Textile house-furnishings | .55 | .56 |
| Furniture and bedding | 1.40 | 1.41 |
| Floor coverings | .39 | .39 |
| Appliances | 1.07 | 1.10 |
| Other house-furnishings | .79 | .79 |
| Housekeeping supplies | 1.36 | 1.35 |
| Housekeeping services | 1.85 | 1.79 |
| Apparel and upkeep | 10.45 | 10.57 |
| Men's and boys' | 2.82 | 2.87 |
| Women's and girls' | 4.02 | 4.06 |
| Footwear | 1.57 | 1.58 |
| Other apparel | 2.04 | 2.06 |
| Transportation | 13.27 | 13.53 |
| Private | 11.80 | 12.08 |
| Autos and related goods | 7.70 | 8.06 |
| Auto services | 4.10 | 4.02 |
| Public | 1.47 | 1.45 |
| Health and recreation | 19.87 | 19.73 |
| Medical care | 6.46 | 6.37 |
| Personal care | 2.58 | 2.60 |
| Reading and recreation | 5.71 | 5.66 |
| Other goods and services | 5.12 | 5.10 |
| Tobacco products | 2.09 | 2.07 |
| Alcoholic beverages | 2.50 | 2.50 |
| Personal expenses | .53 | .53 |
| Miscellaneous | .38 | .38 |

Source: U.S. Department of Labor, Bureau of Statistics.

Also included in table 1 are the relative weights of each item or service. For example, food is weighted 22.19, dairy products 2.72, and medical care 6.46 in December 1971. The weights are not supposed to represent the portion of an eligible family's income that goes for each item or service. This can readily be seen by observing that both the cost of owning a home and of renting a home are included in the category of shelter. No family clearly does both. These weights are revised in December of each year and were initially set by a survey of consumer buying habits. The weights are intended to reflect the amount that the total group of clerical workers and city wage earners spend on each item or service.

The prices of goods and services are collected in 56 cities (39 Standard Metropolitan Statistical Areas and 17 smaller cities) by personal visits to a representative sample of nearly 18,000 stores and service establishments where wage earners and clerical workers buy good and services. Rental rates are obtained through about 40,000 tenants. The cities sampled are given in table 2. An index is calculated for each of these cities and then combined into the CPI of all cities by weighting the individual indexes by the proportion of the eligible population in each city.

**TABLE 2    Cities and Population Weights for CPI**

| City and Size Stratum | Population Weight |
|---|---|
| A. Standard metropolitan statistical areas of 1,400,000 or more in 1960: | |
| Baltimore, Md. | 1.402 |
| Boston, Mass. | 1.930 |
| Chicago-Northwestern Indiana | 5.552 |
| Cleveland, Ohio | 1.325 |
| Detroit, Mich. | 2.895 |
| Los Angeles-Long Beach, Calif. | 5.017 |
| New York-Northeastern New Jersey | 12.577 |
| Philadelphia, Pa. | 2.703 |
| Pittsburgh, Pa. | 1.565 |
| St. Louis, Mo. | 1.428 |
| San Francisco-Oakland, Calif. | 2.372 |
| Washington, D.C. | 1.255 |
| B. Standard metropolitan statistical areas of 250,000 to 1,399,999 in 1960: | |
| Atlanta, Ga. | 2.934 |
| Buffalo, N.Y. | 2.347 |
| Cincinnati, Ohio-Ky. | .740 |
| Dallas, Tex. | 2.934 |
| Dayton, Ohio | 1.096 |
| Denver, Colo. | 1.838 |
| Hartford, Conn. | 2.348 |
| Honolulu, Hawaii | .354 |
| Houston, Tex. | .999 |
| Indianapolis, Ind. | 1.095 |
| Kansas City, Mo.-Kans. | .710 |
| Milwaukee, Wis. | .850 |
| Minneapolis-St. Paul, Minn. | 1.042 |
| Nashville, Tenn. | 2.933 |

**TABLE 2** — *Cont'd.*

| City and Size Stratum | Population Weight |
|---|---|
| San Diego, Calif. | .872 |
| Seattle, Wash. | 1.837 |
| Wichita, Kans. | 1.006 |
| C. Standard metropolitan statistical areas of 50,000 to 249,999 in 1960: | |
| Austin, Tex. | 1.250 |
| Bakersfield, Calif. | 1.323 |
| Baton Rouge, La. | 1.250 |
| Cedar Rapids, Iowa | 1.284 |
| Champaign-Urbana, Ill. | 1.284 |
| Durham, N.C. | 1.250 |
| Green Bay, Wis. | 1.284 |
| Lancaster, Pa. | 1.803 |
| Orlando, Fla. | 1.250 |
| Portland, Maine | 1.803 |
| D. Urban places of 2,500 to 49,999 in 1960: | |
| Anchorage, Alaska | .005 |
| Crookston, Minn. | 1.352 |
| Devils Lake, N. Dak. | 1.352 |
| Findlay, Ohio | 1.352 |
| Florence, Ala. | 1.227 |
| Kingston, N.Y. | 1.171 |
| Klamath Falls, Oreg. | 1.338 |
| Logansport, Ind. | 1.352 |
| Mangum, Okla. | 1.226 |
| Martinsville, Va. | 1.227 |
| McAllen, Tex. | 1.227 |
| Millville, N.J. | 1.171 |
| Niles, Mich. | 1.351 |
| Orem, Utah | 1.339 |
| Southbridge, Mass. | 1.170 |
| Union, S.C. | 1.227 |
| Vicksburg, Miss. | 1.226 |

Source: U.S. Department of Labor, Bureau of Labor Statistics.

## Price Changes versus Quality Changes

It is important to realize that the CPI, as other price indexes, is designed to measure changes in prices of goods and services that experience no change in **quality.** Many critics of the CPI claim that it does not properly take into account quality changes in goods and services that are priced. Some, on the other hand, feel that unmeasured quality improvement probably offsets unmeasured quality deterioration. For example, what part of an increase in the cost of medical services reflects a quality increase, and what part is a real price change? Such a problem has recently been compounded in view of great advances made in medical research. (See Discussion 1.)

The following example should help illustrate how and why we must continually guard against increased political influences on federal statistical agencies.

> To illustrate the dangers of political pressure on statistical decisions, consider the technical problem associated with assigning the cost of air pollution and emission control equipment on automobiles as a component of the Consumer Price Index. There was considerable debate whether to classify this equipment as a quality improvement — consequently, not influencing the Consumer Price Index — or as a cost increase which would be reflected in the Consumer Price Index.
>
> A statistical decision on cost versus quality in automobile pricing has to be made annually and in 1972 it had to be made during an election campaign. If political considerations were to enter this statistical issue, it would be beneficial to labor to include the emission control equipment as a cost increase, thereby adding a "cost-of-living" increase to the wages of millions of workers, and perhaps, politically reflecting adversely on the success of (price) controls in holding down inflation.
>
> Alternatively, political advocates who are concerned with demonstrating the success of anti-inflationary policies would urge classification of this equipment as a quality improvement, as would those interested in demonstrating the increased productivity of labor and the greater output of the economy.
>
> A technical committee of professional statisticians was convened to resolve this statistical issue, and there is no evidence that political pressure was exercised. However, the nature of this type of decision illustrates the importance of producing technical statistical decisions which are above suspicion and maintaining them in an arena which is independent from political pressure.[3]

Assessment of quality is the most difficult problem encountered in a price index such as the CPI. Many decisions must be made concerning whether a change in price was accompanied by a change in quality. If quality changed with price, the CPI will not reflect the full price change. The CPI measures only a price change that is unaccompanied by a quality change.

Change in *quantity* is easily accounted for in price indexes such as the CPI. Prices per unit are recorded so that a change in size of the contents of a package will be reflected in a price change per unit of the item. Hence, reductions in the size of a candy bar would, if this item were priced in the CPI, show increases in the price per ounce.

### Rate of Change

We must be careful when discussing monthly changes in the CPI. As previously mentioned, we are usually concerned with the rate of change of the CPI. At this monthly rate, how much will the CPI change in a year? Rate of change of the index and changes in index points are two distinct concepts. For example, in August 1973 the CPI increased 1.9 index points, from 133.2 to 135.1 with respect to the base year 1967. The rate of change of the index is

$$\frac{(135.1 - 133.2)}{133.2} \times 100 = \frac{190}{133.2} = 1.4\%$$

The prices of the goods and services measured by the CPI increased 1.4% in August of 1973 or at the annual rate of nearly

$$1.4\% \times 12 = 17\%$$

Recall our discussion of figure 3, chapter 3, which is a graph of the rate of change of the CPI, not a graph of the change in index points.

With this brief introduction to the CPI, we should better understand the "billion-dollar index." Keep in mind that this index measures price changes for goods and services purchased by a certain portion of our population. It is not a measure of the cost of living for a family.

## Real, or Constant, Dollars

The media gives much attention to real, or constant, dollars. Constant dollars are monetary amounts reflecting a fixed buying power. Since the buying power of money changes over time, it does not make sense to directly compare dollar amounts for two different time periods. We alluded to this in Discussion 3 in chapter 9. In that case dollar amounts for the years 1937, 1939, and 1951 were compared. Dollar amounts should be reduced to monetary amounts reflecting a constant buying power for meaningful comparisons.

Let us look at an example. Suppose someone's salary changed from $11,500 in 1971 to $14,500 in 1975. How should we interpret this change? Using a price index, let's say the CPI, we can reduce these values to constant dollars with respect to the base year for the CPI, that is, with respect to 1967 dollars. The CPI was 120 in 1971 and 160 in 1975 (approximate September figures). Changed to price relatives with respect to 1967 dollars, the salaries would be

$$\frac{\$11,500}{120} \times 100 = \$9,583.33 \text{ and } \frac{\$14,500}{160} \times 100 = \$9,062.50$$

for 1971 and 1975, respectively; that is, in constant (1967) dollars the 1975 salary is less than the 1971 salary. In fact, these figures represent a 5.4% decrease in salary:

$$\frac{\$9,062.50 - \$9,583.33}{\$9,583.33} \times 100 = -5.4\%$$

The purchasing power of the 1975 salary of $14,500 is less than the purchasing power of the 1971 salary of $11,500.

# Other Index Numbers

It would be a difficult task indeed to try and discuss all of the index numbers we are likely to encounter. There are however a few important index numbers that warrant mention.

## *Wholesale Price Index*

The oldest continuous statistical series published by the Bureau of Labor Statistics is the Wholesale Price Index (WPI), which was first published in 1902. Since its beginning, the WPI has become increasingly representative of primary market price changes. The word "wholesale" used in the title refers to sales of large quantities, not to prices received by wholesalers or distributors. The WPI measures the change in prices of items the first time they are sold. For example, the WPI measures the first sale of raw materials such as those listed in table 3 as well as the first sale of any product made from the raw materials.

A glance at table 3 gives some indication of the products and materials priced in the WPI.

TABLE 3    Major Groups of Items Priced in WPI

Farm products
Hides, skins, leather, and related products
Fuels and related products, and power
Chemicals and allied products
Processed foods and feeds
Lumber and wood products
Metals and metal products
Machinery and equipment
Drugs and pharmaceuticals
Rubber and plastic products
Nonmetallic mineral products (glass, asphalt, etc.)
Electrical machinery and equipment
Furniture and household durables (including transportation equipment such as cars)
Miscellaneous products (such as toys, sporting goods, tobacco products and photographic equipment)

Source: U.S. Department of Labor, Bureau of Labor Statistics.

Although the WPI may indicate an approaching change in retail prices, it does not directly reflect a change in consumer prices. The index is generally used in long-term contracts to escalate material prices; it is also used in budget making, in planning cost of plant expansion, in appraising inventories, and in establishing replacement costs.[4]

The WPI is of direct use to us for indicating what is to come as far as consumer or retail prices are concerned. Thus, the direction of the CPI may be forecast to some extent by the movement of the WPI.

## Crime Rates

As mentioned at the beginning of this chapter, the simplest type of index number is a percentage change such as the percentage increase or decrease of the crime rate in the United States. The crime rate, a part of the FBI's *Uniform Crime Report*, is the number of reported major offenses per 100,000 population. This statistic is used to demonstrate the risk of becoming a victim of a crime. Since any citizen of our country could be a victim of a crime, the crime rate reflects the number of reported crimes per 100,000 citizens. The major crimes included in the crime rate are murder, forcible rape, robbery, aggravated assault, burglary, larceny, and auto theft.

Observe that crime is defined as reported crime. The first link in the reporting chain is the victim of crime, who must notify law enforcement officials. Many crimes, notably rape, are never reported to the authorities and are therefore not accurately reflected in the crime rate. Also, note that crime as defined in this index has nothing to do with whether anyone was arrested for (or convicted of) the crime in question. (Recall from our discussion in chapter 2 that knowing definitions is important to proper interpretation of a statistic.)

The source of raw data on reported crime is often the local police departments, which may report directly to the FBI. A few states accumulate crime information at the state level. These data are then sent to the FBI. In states where data are collected at the state level, duplication of reports (a measurement problem) can be checked. Duplication may result when two police departments (for example, city and county) may have the same crime reported to them. In states not having a centralized crime statistics agency, the local law enforcement groups report directly to the FBI. The possibility of duplication and other types of inaccuracies is obvious, especially since the crime rate of certain cities is a sensitive political issue.

The reporting of the crime rate is accompanied by the percentage change, the rate of change, based on the preceding reporting period. Since the crime rate is the proportion of reported crimes per

100,000 population, the rate may decrease although the number of reported crimes actually increases. This happens when the population increases faster than the increase in reported crimes.

A complete list of index numbers would be endless. A few noteworthy indexes are: the Gross National Product (GNP), which represents the purchase of goods and services by consumers and government, domestic investment, and exports; the real spendable earnings index, which is an adjustment of earnings data according to federal income and Social Security taxes; the farm parity price indexes; and the employment/unemployment figures. The interpretation of certain index numbers will be discussed in the next section.

## Interpreting Index Numbers

Let us consider some important aspects of interpretation of index numbers by discussing how we might interpret the Consumer Price Index and the FBI's *Uniform Crime Report* (*UCR*). If we can understand how to view changes in the CPI and the crime rate, then we should have a good understanding of the many different index numbers our government leaders use in making decisions. Such an interpretation will, of course, require a knowledge of what the particular index is supposed to measure. This was discussed in the last two sections.

### CPI

Recall that the CPI is first calculated in 56 metropolitan areas before being combined into a national index. It would be very informative to know the city index for your place of residence, if you live in one of the 56 metropolitan areas sampled.

The Bureau of Labor Statistics publishes both seasonally adjusted and unadjusted changes for the CPI. For example, in the food component of the CPI certain produce items have prices that rise and fall in a periodic pattern according to growing seasons. The price of vegetables is high until the harvest, when greater supply forces down the price. Seasonally adjusted figures will not reflect price changes that are the result of such a predictable price pattern.

> For analyzing general price trends in the economy, seasonally adjusted changes are usually preferred since they eliminate the effect of changes that normally occur at the same time and in about the same magnitude every year — such as price movements resulting from changing climactic conditions, production cycles, model changeovers.

The unadjusted data are of primary interest to consumers concerned about the prices they actually pay. Unadjusted data are also used extensively for escalation purposes. Many collective bargaining contract agreements and pension plans, for example, tie compensation changes to the CPI unadjusted for seasonal variation.[5]

In summary, recall our earlier consideration in this chapter of what the CPI does measure, and remember that it does not measure changes in our cost of living. Look for CPI figures for your city or for a city of comparable size and location. Note whether reported figures are adjusted seasonally or not. Also, recall that earlier in this chapter we stressed that you should observe whether a reported figure represents a change in index points or a rate of change of the index.

## Crime Rates

Switching to crime statistics, it should be emphasized that governmental statistics reflect national trends, not local changes, which would interest us most. As an example, suppose you are a resident of the SMSA of Chicago, Illinois. The 1976 UCR indicates that the national crime rate was 5,266.4. (Recall that the crime rate is the number of reported major crimes per 100,000 population.) But this figure is, as indicated in the title, **Uniform Crime Report**, uniform for the whole country. It reflects the Nevada rate of 8,306.1 as well as West Virginia's 2,319.7 rate. Hence, for a resident of Chicago the national crime rate is not very descriptive. The state of Illinois reported a 5,055.0 crime rate for 1976. This is more pertinent information for a Chicagoan, but the Illinois rate is still a report of all crime in the state. From the UCR we note that the 1976 Chicago crime rate was 5,756.0. This information is disheartening but more informative than the crime rate for the United States or Illinois.

When contemplating buying a house in Chicago, one must be concerned about crime in the area of Chicago. Different areas within the SMSA of Chicago (Cook, DuPage, Kane, Lake, McHenry, and Will Counties) will have crime rates different from the total rate for Chicago.

The purpose of the preceding example is not to scare you by throwing around so many statistics. The point is this — governmental indexes reflect national trends and will therefore be used mostly by those who determine national policy. If you are interested in what is happening to prices, crime, unemployment, etc. in a particular area, national statistics will be of little use. You must try to find state or local statistics to guide you. There are various sources of such information. A local newspaper is likely to have the data, although these data may or may not have been published. Otherwise, the best source of the information is the government agency that reports the data. For example, the CPI, WPI, and employment statistics are published by the Bureau of Labor Statistics, and crime rates are made available through the FBI.

# Summary

Index numbers are statistics that reduce data to values relative to a particular quantity or base. We have considered both the calculation and interpretation of these statistics.

The Consumer Price Index is not a measure of a family's cost of living. It is a measure of price changes of goods and services bought by urban wage earners and clerical workers as a group.

We must be careful when reading about the CPI in the news media. Statistics reported can either be the rate of change of the CPI (usually seasonally adjusted) or the change in index points (adjusted or unadjusted). Often reports on this index number are not clear as to which statistic was used.

Price indexes are a measurement of price changes unaccompanied by a quality change. Also, note that index numbers describe national economic trends, not the changes you will necessarily encounter where you live.

# No Comment

■ *"Some statistics may be inaccurate, but those that deal with the cost of living are on the up and up."*
(Lexington Herald, *31 December 1976*)

■ *"These statistics are awful," he (George Meany, president of the AFL-CIO) said. "They come as no surprise to the worker and consumer who needed no additional evidence [of a decrease in buying power of pay checks]."*

# Exercises

1. What does it mean when a newspaper reports that one third of the average increase of $316 per car in the price of new model cars is traceable to improvement and two-thirds is a plain price increase?

2.    What are some of the problems involved in pricing food items in the many different food stores in the 56 different cities considered in the CPI of all cities combined?

3.    Why is it desirable that a revised CPI include segments of our population other than clerical workers and urban wage earners?

4.    How would you define "crime"? Is your definition "workable"? Is the FBI's definition of crime in the *Uniform Crime Report* "workable"?

5.    Investigate the report of a change in the CPI in your local newspaper(s). Is it clear whether the statistics mentioned indicate a change in index points or a rate of change? Are the figures seasonally adjusted?

6.    What seasonal adjustments might be made in employment/unemployment figures?

7.    The description of urban wage earners and clerical workers in this chapter included reports of mean salaries, mean family size, etc. Recalling our discussion in chapter 3 on indexes of central tendency, would you consider that description to be a good way to describe this group of people?

8.    Consider the prices listed in table 4, which gives the mean price per unit for a number of stores in a metropolitan area.
      What is the percentage increase in the price of sugar, bread, coffee, and cereal from July 1972 to April 1974? Including the relative importance of these items, what is the percentage increase in the price of all 4 items over this two-year period?

TABLE 4    Prices of Goods

|        | July 1972 | April 1974 | Relative Importance |
|--------|-----------|------------|---------------------|
| Sugar  | $.659     | $1.193     | .57                 |
| Bread  | .287      | .476       | .55                 |
| Coffee | .905      | 1.233      | .40                 |
| Cereal | .301      | .379       | .74                 |

# For Discussion

1. Discuss the change in *quality* of medical services. How does this affect the medical care component of the CPI, which has been steadily increasing over the years?

2. Should Social Security benefits be tied to the CPI as it was designed prior to 28 February 1978? For that matter, should any worker's wages be so tied?

3. Certain candy companies have been using a chocolate substitute — a substance derived from cotton plants. If the chocolate candy bars of these companies are priced in the CPI, should a decrease in quality be reflected by a price increase? Discuss the general problem of price indexes that are intended to measure price changes that are unaccompanied by a quality change.

# Further Readings

For more details concerning the Consumer Price Index you should refer to the articles listed in the Notes section of this chapter.

If you would like to take a more general look at economic statistics, I suggest the article "Early Warning Signals for the Economy" by G. Moore and J. Shiskin in Tanur's *Statistics: A Guide to the Unknown* (San Francisco: Holden-Day, 1972). This article discusses leading and lagging economic indicators. Leading indicators precede (lead) a change in the economy, that is, these indicators predict that a change is coming. Lagging indicators follow an economic change.

# Notes

1. Arthur Sackley, "Coinage, Commodities and Count Carli: An Account of the Inventor and the Computation of the Original Index Numbers," *Monthly Labor Review* (July 1965): 818.
2. U.S. Department of Labor, Bureau of Labor Statistics, "The CPI: A Short Description," Bulletin 0-265-848, 1967, p.1.
3. "Maintaining the Professional Integrity of Federal Statistics," *American Statistician*, vol. 27, no. 2, pp. 63-64.
4. U.S. Department of Labor, Bureau of Labor Statistics, *Handbook of Methods*, Bulletin 1711, rev. 1972, p. 105.
5. U.S. Department of Labor, Bureau of Labor Statistics, *The Consumer Price Index* (March 1972): 22.

# 11 Doublespeak Revisited

WE DISCUSSED STATISTICAL DOUBLESPEAK in the first section of this book. We now return to this concept and ask whether the knowledge acquired through our discussions of statistics will immunize us from the statistical doublespeak of public officials and advertisers.

## Statistical Doublespeak

In the first chapter of this book we discussed the formation of the Committee on Public Doublespeak by the National Council of Teachers of English (NCTE). This committee was created to combat "the dishonest and inhumane uses of language and literature of advertisers" and the "semantic distortion [of language] by public officials." Doublespeak was defined as "inflated, involved and often deliberately ambiguous language."

It is difficult to think of any language that is more involved, more inflated, and more ambiguous than the language of statistics. Statistics is a common type of doublespeak primarily because people just do not understand this language; indeed, many admit to being afraid of those mysterious numbers called statistics.

It is hoped that after a careful journey through this book you have been able to break through the barrier that likely once separated you from "those numbers." You are, I hope, at the point where you will read and think about statistics. You may even have become quite good at detecting the great amount of statistical doublespeak that pervades our media.

But an understanding of statistics, though necessary, is not enough. We all understand the English language, yet we all are susceptible to the verbal doublespeak of public officials and advertisers. Hence, understanding statistics will not by itself protect us against statistical doublespeak. Let us look at the anatomy of public doublespeak so you may be better able to assimilate the propaganda fed us daily by the media.

## The Anatomy of Doublespeak

Hugh Rank, first chairperson of the NCTE Committee on Public Doublespeak, has developed a schema (diagram) for public doublespeak, which is essentially an anatomy of doublespeak. This schema is shown in figure 1.

The communication of public officials and advertisers can result in doublespeak by intensifying the good and downplaying the bad. This is just as true with statistical doublespeak as it is with verbal double-

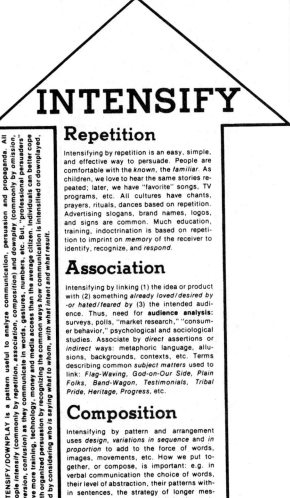

INTENSIFY/DOWNPLAY is a pattern useful to analyze communication, persuasion and propaganda. All people intensify (commonly by repetition, association, composition) and downplay (commonly by omission, diversion, confusion) as they communicate in words, gestures, numbers, etc. But, "professional persuaders" have more training, technology, money and media access than the average citizen. Individuals can better cope with organized persuasion by recognizing the common ways how communication is intensified or downplayed, and by considering who is saying what to whom, with what intent and what result.

## Repetition

Intensifying by repetition is an easy, simple, and effective way to persuade. People are comfortable with the *known*, the *familiar*. As children, we love to hear the same stories repeated; later, we have "favorite" songs, TV programs, etc. All cultures have chants, prayers, rituals, dances based on repetition. Advertising slogans, brand names, logos, and signs are common. Much education, training, indoctrination is based on repetition to imprint on *memory* of the receiver to identify, recognize, and *respond*.

## Association

Intensifying by linking (1) the idea or product with (2) something *already loved/desired by* -or *hated/feared by* (3) the intended audience. Thus, need for **audience analysis**: surveys, polls, "market research," "consumer behavior," psychological and sociological studies. Associate by *direct* assertions or *indirect* ways: metaphoric language, allusions, backgrounds, contexts, etc. Terms describing common *subject matters* used to link: Flag-Waving, God-on-Our Side, Plain Folks, Band-Wagon, Testimonials, Tribal Pride, Heritage, Progress, etc.

## Composition

Intensifying by pattern and arrangement uses *design*, *variations in sequence* and *in proportion* to add to the force of words, images, movements, etc. How we put together, or compose, is important: e.g. in verbal communication the choice of words, their level of abstraction, their patterns within sentences, the strategy of longer messages. **Logic**, inductive and deductive, puts ideas together systematically. **Non-verbal** compositions involve *visuals* (color, shape, size); *aural* (music); *mathematics* (quantities, relationships) *time* and *space* patterns.

**FIGURE 1**   The Rank Schema of Public Doublespeak

Source: © 1976 by Hugh Rank, Permission to reprint for educational purposes hereby granted, *pro bono publico*. Endorsed by the Committee on Public Doublespeak, National Council of Teachers of English (NCTE).

**FIGURE 1 —** *Cont'd*

## Omission

Downplaying by omission is common since the basic selection/omission process *necessarily omits* more than can be presented. All communication is limited, is edited, is slanted or biased to include and exclude items. But omission can also be used as a *deliberate* way of concealing, hiding. Half-truths, quotes out of context, etc. are very hard to detect or find. Political examples include *cover-ups, censorship, book-burning, managed news, secret police activities.* Receivers, too, can omit: can "filter out" or be closed minded, prejudiced.

## Diversion

Downplaying by distracting focus, diverting attention away from key issues or important things; usually by intensifying the side-issues, the non-related, the trivial. Common variations include: *"hairsplitting," "nit-picking," "attacking a straw man," "red herring";* also, those emotional attacks and appeals (*ad hominem, ad populum*), plus things which drain the energy of others: *"busy work," legal harassment,* etc. Humor and entertainment (*"bread and circuses"*) are used as pleasant ways to divert attention from major issues.

## Confusion

Downplaying issues by making things so complex, so chaotic, that people "give up," get weary, "overloaded." This is dangerous when people are unable to understand, comprehend, or make reasonable decisions. Chaos can be the accidental result of a disorganized mind, or the deliberate flim-flam of a *con man,* or the political *demagogue* (who then offers a "simple solution" to the confused.) Confusion can result from *faulty logic, equivocation, circumlocution, contradictions, multiple diversions, inconsistencies, jargon* or anything which blurs clarity or understanding.

# DOWNPLAY

speak. For example, cigarette and alcohol ads intensify the pleasure of smoking and drinking, downplaying the health hazards, Loan companies intensify the low down payment, downplaying the high interest rate and the term of the loan. Ads for products will generally, intensify the quality of the product and downplay the quantity or cost.

In the case of statistical doublespeak a number may be meaningless and thus should be discounted. A credible statistic, on the other hand, has its limitations. For example, the sample from a well-designed and well-conducted survey is only representative of the population actually sampled. You should be able to properly interpret survey (and other) statistics, credible or not, after our discussions in this book.

## Techniques for Intensifying

Regardless of whether a statistic is credible, public officials and advertisers can still be guilty of statistical doublespeak. For example, they can intensify or overemphasize the good qualities of their statistics through repetition, association, and composition, as Rank suggests. Let us look at each of these techniques.

**Repetition**   Repeat those statistics that are favorable. Advertisers are effective in this area. How many times have we heard, "4 out of 5 dentists surveyed recommend a particular sugarless gum for their patients who chew gum?" We hear a claim like this so often that the phrase "4 out of 5 dentists" is ingrained in the mind. We are likely to forget that many dentists probably don't want any of their patients to chew any type of gum. (This statistic is not mentioned — downplay by omission.) We soon would not even care about the accuracy of the "survey." So, even when a statistic is of questionable importance, we become so comfortable with the number by its repetitive use that it is no longer questioned.

**Association**   Statistics are associated with scientific "truth." Such catch phrases as, "Statistics prove" or, "But statistics say" are commonly used in this regard. Intensifying by association is particularly effective if people are afraid of statistics or think of statistics as being gospel and the last word.

**Composition**   The level of abstraction, or confusion, is a key to the use of statistics for intensifying by composition. People don't question statistics as the accompanying explanations are themselves confusing. These explanations are themselves doublespeak since few listeners will understand or even care. Do not be intimidated. You are likely to be just as versed in statistics as the person trying to overwhelm you.

## Techniques for Downplaying

Public officials and advertisers not only intensify the good statistics, but they will downplay the statistics that are not favorable to them through omission, diversion, or confusion. Once again, these statistics may or may not be credible.

**Omission**    Advertisers just do not mention unfavorable statistics. Advertisements report (intensify) only the favorable statistics. For example, recall the example of the chewing gum ad. As another example, consider the public officials who talk about the low inflation rate although the high unemployment figures are not mentioned. Many statistics are available on the state of our economy: some may be good, and others reflect poorly on the people in power. Rather than look at the overall picture, a public official will likely intensify the favorable data while omitting the unfavorable.

**Diversion**    This is so easy with statistics. How many times have you heard a statement like, "Statistics? I don't trust them. You can make them say whatever you want." Divert attention from what statistics should be telling people by attacking statistics in general. For example, a teacher claims that teachers should not be held accountable for their performance in the classroom since that means another statistic (measurement of effectiveness), and, "There are too many statistics around." Attack statistics in general, drawing attention away from the real issue.

**Confusion**    Downplay by making things complex and chaotic. How easy it is to confuse people by using statistics! Here's where we need to be very careful. However, it is hoped that statistics are a little less confusing than they were before we started our discussions. Don't let a little knowledge hurt you. Be on the offensive. Do not allow someone to confuse you — challenge them. They will likely become confused. The next time someone tries to cloud an issue with a statistic, challenge that person. You are likely to find that people are not able to defend their statistics. If they can defend the statistics — fine: then, there is important information contained in the statistic.

   If you would like to analyze a statement made by a public official or advertiser, whether or not the statement is statistical, ask yourself what is not being said. Don't be diverted from finding the information you desire, and be on the offensive lest you become confused.

   Be on guard against attempts to intensify the good. When you know what is going to be said even before it is said, question the repetition. Judge what is being said not how it is being communicated.

   Remember that a politician and an advertiser are trying to influence you, but there are a rival political candidate and a competing product. What one side is intensifying, the other side is downplaying. Listen to both sides as each will point out the weaknesses of the other. Then decide.

# Summary

Throughout this book we have tried to break the barrier that most likely prevented you from even thinking about statistics. It is likely that you now think about the statistics you read and hear in the media. You may even be quite good at separating the meaningless numbers from the meaningful statistics.

But knowing when a number is a meaningful statistic is not enough. Knowing the limitations of the statistic is not even enough. No one, no matter how much knowledge of statistics he or she possesses, is immune from the statistical doublespeak of public officials and advertisers. Be on guard against public officials and advertisers who will downplay the statistics (credible or not) that reflect poorly on them and intensify the numbers that are favorable. In fact, be on guard against both the verbal and statistical doublespeak of public officials and advertisers.

By using the ideas and techniques presented in this book, you should be able to evaluate the doublespeak you hear and read and sort out the meaningless information from the credible information.

# No Comment

■ *Concerning the proposal of the National Council of Teachers of English to establish the Committee on Public Doublespeak:*

> *"The proposal, which seems to us to represent relevant teaching at its best, may not find itself overwhelmed with offers for Federal aid. Few members of Congress are likely to run to introduce bills for the subsidy of truth in rhetoric. But if the nation's English teachers support their colleagues' drive against linguistic pollution, they will be acting in the spirit of Thomas Jefferson's statement of faith: 'Enlighten the people generally, and tyranny and oppression of body and mind will vanish like evil spirits at the dawn of day.' "*
> (New York Times *editorial, 12 December 1973)*[1]

# For Discussion

Listen to and read the statements of public officials and advertisers. Look for examples of intensifying favorable statistics while downplaying those that are unfavorable. Also look for examples of verbal doublespeak.

# Further
# Readings

You just finished reading the only book I know of that is written on the subject of statistical doublespeak. On the subject of verbal doublespeak I think you'll find the books by Rank and Dieterich quite enjoyable. You will recall we mentioned Hugh Rank was the first chairperson of the NCTE's Committee on Public Doublespeak and Daniel Dieterich is the second chairperson. Their books are *Language and Public Policy* (1974), H. Rank editor, and *Teaching Public Doublespeak* (1976), D. Dieterich editor. Both books are published by the National Council of Teachers of English, Urbana, Illinois.

# Notes

1. © 1973 by The New York Times Company.

# Appendix A:
# Two Interviews

## A Gallup Interview

YOU CAN READ ABOUT national polling organizations such as Gallup's American Institute of Public Opinion or even read about polling in general, as in this book; but actually experiencing an interview by a Gallup representative is an interesting experience. I know — it happened to me.

In May of 1975 I was interviewed for a Gallup survey and would like to share this interesting experience with you. The interview itself lasted about 30 minutes. Questions covered such diverse topics as abortion, luggage, banks and banking, wines, microwave ovens, potential candidates for the upcoming presidential election (a year and a half away), advertising in the media, corporate relationships to consumers, lawn mowers, and sinus congestion as well as the usual demographic questions on age, income, education, etc. Quite clearly, you have little time to do much thinking about the questions asked, since you are continually being asked to change thought processes.

More interesting to me and more relevant to our discussion was *my interview of the Gallup interviewer.*

The interviewer was a young woman who worked for an interviewing service located in a nearby metropolitan area. She was paid $2.25 per hour and 12¢ a mile for her work. For her first two Gallup surveys the interviewer received instructions from the interviewing organization for which she worked. After completing the first two Gallup surveys, she received questionnaires and necessary instructions directly through the mail.

The interviewer had been asking questions for a different Gallup survey each week for the last 10 or 11 months! Each survey was conducted on a different block in the same general area of our city; that is, each week, usually on Saturday, she would ask a different set of questions (a different Gallup survey) of a member of each household of a number of households on different blocks in our area of town.

Her instructions were to interview 5 adults on a block for each week's survey. One week she'd talk to three males and two females, the next week two males and three females. Occasionally she would be instructed to talk to a special sixth person, for example, a male 16 years of age.

She was instructed to start at a particular corner of our block, the east corner, go to the fifth household and then to every household until her quota of 5 interviews was filled. A household is defined as a "living quarters" for a family or individual(s). The building at the east corner of our block houses 4 apartments, so the fifth household was the resident in the next house (going clockwise).

Each household visited would be classified as "not home," "refused to be interviewed," "completed interview," etc. The interviewer I talked to said that she would have to contact only 6 or 7 adults to get the required 5 interviews. My interviewer indicated that she was quite good at talking people into being interviewed and this is why she required only 6 or 7 contacts. According to her, some less experienced interviewers may have a harder time getting 5 completed interviews. (If an interviewer had visited all households on a block, he or she would have to revisit households until 5 interviews were completed.)

I was informed that it was not unusual to be "checked." Since one of the questions asked was my phone number, I was told that someone might call to confirm that the interview had indeed taken place.

There are other reasons for getting my phone number, as I found out a few months later. A different Gallup representative called to conduct a second interview. This telephone interview was 20 minutes long and involved questions only about wine!

My second interviewer was a male who was making his first series of interviews. He had answered an ad in a local newspaper, had taken a test (a trial interview), and then waited until he was given his present assignment.

There is yet another way to use telephone numbers acquired by means of a personal interview survey: the numbers can form a basis for a subsequent telephone survey. If a person interviewed has a telephone number of, let's say, 296-5280, a second survey may be conducted by telephone using a number beginning 296-52 — and randomly selecting the last two digits from a table of random numbers. The first 5 digits are held fixed since, if these numbers are assigned, all such numbers are likely to be assigned. Also, telephone numbers with the same first 5 digits are likely to be in the same area of a telephone district.

## An Arbitron Diary

I have heard many people comment that they have never been asked their opinions for a survey and therefore think that survey results do not reflect their feelings properly. Yet, within a year of being interviewed by the Gallup organization, I was asked to participate in an Arbitron radio listening survey.

Arbitron (American Research Bureau) conducts both television and radio surveys in local markets. Local survey markets are generally an enlargement of a Standard Metropolitan Statistic Area. The total survey area consists of the SMSA plus contiguous counties that receive strong (radio or television) signals from stations in the metropolitan area.

One of the important measurements made by television and radio survey groups is the audience share. The *audience share* for a radio station is the *proportion of people listening* to a radio at a particular time who are listening to a particular station. The audience share for a television program is defined similarly. (Would there be interest in the number of people viewing at a particular time?)

A letter was sent to my home by Arbitron indicating that we would be called and asked to participate in a radio survey. This premailing was to assure us that no one was trying to sell anything. (A premailing of this type is not uncommon, since some salesmen start a sales pitch with, "I'm conducting a survey." People will consequently refuse to talk to a representative of a legitimate survey organization for fear that halfway through the "questions" they will hear something like, "Oh, your car is 4 years old; would you be interested in a new car?" Licensing of surveys is consequently required in some cities.)

I was subsequently contacted by telephone. After agreeing to participate, a diary was sent each member of our household who is over 12 years of age. (For a television survey a diary is sent for each television set.) Enclosed with a letter of appreciation was a token of 50¢ for each participant from our household.

Before the survey week began we were again contacted. This call verified the arrival of the diaries and reminded us to keep an accurate record of our listening habits for the survey week, 15 April to 21 April. On the next to the last day of our survey week we were contacted and reminded to mail in our completed diaries on 22 April.

I found it quite difficult to remember to record in the diary each and every time I listened to the radio: sometimes at home, other times in the car, or even at a business. (A car dealer must have thought me strange when I asked what radio station they had coming over their speaker system.)

All in all, the diary was probably a good indication of my radio listening habits for that week considering that:

1. Easter fell on 18 April, and we had out-of-town company most of the week. (Timing of a survey is important.)

2. On the first day of the survey week our television "went on the fritz," so we did not have a working television during the entire week. (We therefore represented people with television sets that were temporarily out-of-order.)

(With regard to our Easter survey week: this particular market is surveyed for radio listening once a year; the survey period is 4 weeks in length although an individual keeps a diary for only one week; the total number of diaries is apparently spread evenly over these 4 weeks; the fact that Easter fell during our week does not mean that all diaries included this holiday.)

How accurate is an Arbitron survey? Consider:

1. Selection of households is from telephone directories except in areas of high concentrations of black or Spanish-speaking people.

2. Diary-type surveys have low response rates. (For a radio survey in our area about 50% of the people contacted apparently returned useable diaries.)

3. The sample size is small. An effective sample size is approximately 600 in an area of about 500,000 people aged 12 or older. (How many are listening to a radio at a particular time?)

The written report of the listening habits in an Arbitron market carefully lists the limitations of the survey. The important question remains unanswered: "How representative of the people 12 years old or older in a particular market area are the survey results?"

Also, recall from our discussions in chapter 2 that the FCC has asked that station contests not be run during survey weeks. Arbitron says that if such a contest is run, a notice will be placed on the report of the survey results. It would be interesting to see this report because one station was running a "cash call" contest (guess a three-digit combination and win up to $1,000 in merchandise).

In this regard, the national networks know when Neilsen surveys are run. Special programming will be aired during these periods. Any comment? For more on Neilsen see M. Wheeler, *Lies, Damn Lies, and Statistics* (New York: Liveright, 1976), Chapter 10.

# Appendix B: Tables

## TABLE 1　Normal Probabilities*

| A | B |
|---|---|
| 3 | 99.7 |
| 2.5 | 99 |
| 2 | 95 |
| 1.645 | 90 |
| 1.5 | 87 |
| 1 | 68 |
| .5 | 38 |

*Within A standard deviations of the mean are B% of data with a normal distribution.

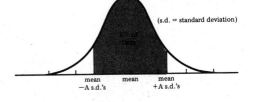

(s.d. = standard deviation)

mean
−A s.d.'s

mean

mean
+A s.d.'s

## TABLE 2　Chi-Square Critical Values

| | Probability of a Greater Value | | |
|---|---|---|---|
| Degrees of Freedom | .1 | .05 | .01 |
| 1 | 2.7 | 3.8 | 6.6 |
| 2 | 4.6 | 6.0 | 9.2 |
| 3 | 6.3 | 7.8 | 11.3 |
| 4 | 7.8 | 9.5 | 13.3 |
| 5 | 9.2 | 11.1 | 15.1 |
| 6 | 10.6 | 12.6 | 16.8 |
| 7 | 12.0 | 14.1 | 18.4 |
| 8 | 13.4 | 15.5 | 20.1 |
| 9 | 14.7 | 16.9 | 21.7 |
| 10 | 16.0 | 18.3 | 23.2 |
| 11 | 17.3 | 19.7 | 24.7 |
| 12 | 18.5 | 21.0 | 26.2 |
| 13 | 19.8 | 22.4 | 27.7 |
| 14 | 21.1 | 23.7 | 29.1 |
| 15 | 22.3 | 25.0 | 30.6 |
| 16 | 23.5 | 26.3 | 32.0 |
| 17 | 24.8 | 27.6 | 33.4 |
| 18 | 26.0 | 28.9 | 34.8 |
| 19 | 27.2 | 30.1 | 36.2 |
| 20 | 28.4 | 31.4 | 37.6 |

**TABLE 3**   **Random Digits**

| Row | 01 05 | 06 10 | 11 15 | 16 20 | 21 25 | 26 30 | 31 35 | 36 40 | 41 45 | 46 50 |
|---|---|---|---|---|---|---|---|---|---|---|
| 01 | 22925 | 76687 | 68317 | 50919 | 87455 | 64464 | 84746 | 35396 | 39889 | 45709 |
| 02 | 23630 | 13318 | 16605 | 22602 | 66245 | 25021 | 61769 | 26246 | 51266 | 93920 |
| 03 | 62402 | 42238 | 93970 | 28206 | 46921 | 05487 | 51983 | 39682 | 11315 | 60751 |
| 04 | 70150 | 88705 | 33143 | 22096 | 16861 | 99224 | 36706 | 49780 | 25410 | 79779 |
| 05 | 13391 | 11614 | 21288 | 78336 | 98293 | 78848 | 36542 | 47690 | 82732 | 10252 |
| 06 | 94326 | 68425 | 72906 | 72203 | 34003 | 41031 | 38462 | 96184 | 48009 | 09651 |
| 07 | 51983 | 76180 | 70444 | 34020 | 46076 | 47552 | 74816 | 65424 | 42187 | 44138 |
| 08 | 77441 | 62579 | 52405 | 09286 | 98474 | 82957 | 19272 | 29159 | 38542 | 46785 |
| 09 | 32038 | 24274 | 80138 | 02570 | 98588 | 90450 | 10552 | 23786 | 25335 | 13426 |
| 10 | 97930 | 62914 | 44576 | 91343 | 27081 | 40802 | 23461 | 64411 | 72328 | 76810 |
| 11 | 19413 | 63275 | 15858 | 59491 | 15807 | 35313 | 05527 | 43006 | 54617 | 49921 |
| 12 | 91554 | 62386 | 63361 | 04586 | 58091 | 17219 | 96997 | 91613 | 40083 | 77775 |
| 13 | 56895 | 42751 | 12485 | 77765 | 04741 | 26089 | 06997 | 78791 | 61491 | 60802 |
| 14 | 41680 | 92583 | 73088 | 06473 | 83470 | 30214 | 15030 | 68157 | 64090 | 98054 |
| 15 | 82634 | 29818 | 97170 | 95858 | 66819 | 25570 | 96793 | 85642 | 41926 | 77214 |
| 16 | 52326 | 06239 | 40218 | 50853 | 15466 | 32106 | 25294 | 96362 | 82100 | 84083 |
| 17 | 59277 | 88094 | 97498 | 86382 | 08365 | 60659 | 57508 | 58533 | 57789 | 25383 |
| 18 | 99485 | 31660 | 77725 | 16513 | 73283 | 49008 | 56760 | 44113 | 59208 | 74896 |
| 19 | 46303 | 99586 | 07241 | 71548 | 29748 | 84064 | 20694 | 97012 | 74680 | 77751 |
| 20 | 86994 | 88574 | 01532 | 70642 | 12975 | 47119 | 92326 | 77053 | 92381 | 09492 |
| 21 | 42979 | 35379 | 55880 | 83396 | 75431 | 08338 | 52910 | 78914 | 39840 | 94810 |
| 22 | 29005 | 08930 | 27447 | 46092 | 20174 | 63395 | 06687 | 38218 | 73951 | 45498 |
| 23 | 69262 | 08253 | 77163 | 74508 | 97194 | 47622 | 26874 | 89696 | 71692 | 74014 |

| 24 | 82436 | 13792 | 01078 | 66156 | 03442 | 60370 | 44407 | 71959 | 25031 | 55381 |
|----|-------|-------|-------|-------|-------|-------|-------|-------|-------|-------|
| 25 | 34952 | 68705 | 56457 | 46206 | 73512 | 21954 | 62743 | 47469 | 77642 | 47673 |
| 26 | 92190 | 82860 | 40526 | 46800 | 65348 | 63598 | 36054 | 39955 | 37115 | 23777 |
| 27 | 92900 | 65253 | 78654 | 14123 | 89352 | 05575 | 92354 | 94142 | 30919 | 26319 |
| 28 | 70077 | 09634 | 48262 | 99507 | 86790 | 39082 | 11788 | 81438 | 96979 | 52653 |
| 29 | 24577 | 82904 | 53996 | 16737 | 18972 | 49198 | 67384 | 95693 | 98629 | 52653 |
| 30 | 46667 | 22174 | 06524 | 29186 | 10915 | 86404 | 06845 | 82326 | 02022 | 69400 |
| 31 | 16603 | 04382 | 63680 | 92215 | 39930 | 90171 | 43597 | 07110 | 62874 | 17503 |
| 32 | 90784 | 59290 | 45438 | 78430 | 44660 | 76410 | 60709 | 29931 | 45454 | 30777 |
| 33 | 84486 | 86877 | 57908 | 80135 | 73361 | 22752 | 73191 | 10822 | 54304 | 70795 |
| 34 | 97486 | 43669 | 04801 | 08724 | 15684 | 54367 | 74409 | 43336 | 01998 | 10254 |
| 35 | 12698 | 41437 | 81159 | 56095 | 44694 | 36083 | 72166 | 37839 | 42084 | 80221 |
| 36 | 57652 | 84335 | 70995 | 54576 | 24579 | 02032 | 75625 | 48777 | 65730 | 35181 |
| 37 | 87006 | 31818 | 65923 | 97656 | 73803 | 68863 | 34191 | 40169 | 24805 | 69555 |
| 38 | 91048 | 62775 | 39874 | 52831 | 03960 | 46915 | 37149 | 92805 | 09714 | 62073 |
| 39 | 43520 | 32558 | 35884 | 92941 | 24829 | 74347 | 23882 | 86247 | 16172 | 33877 |
| 40 | 69781 | 05655 | 31631 | 40936 | 40021 | 09956 | 30328 | 92713 | 58161 | 94020 |
| 41 | 15726 | 59334 | 11128 | 41998 | 68095 | 16239 | 62451 | 70005 | 82640 | 35417 |
| 42 | 82854 | 96060 | 81753 | 48852 | 71341 | 27371 | 78675 | 34929 | 54972 | 05525 |
| 43 | 15848 | 89109 | 48944 | 40981 | 21312 | 47677 | 97805 | 49129 | 97416 | 89941 |
| 44 | 62356 | 08415 | 04305 | 48159 | 26530 | 97420 | 69908 | 75403 | 32118 | 46194 |
| 45 | 18622 | 38570 | 53565 | 55436 | 43099 | 32563 | 65651 | 28162 | 49608 | 62465 |
| 46 | 81083 | 33864 | 59588 | 02302 | 85973 | 82253 | 53367 | 65334 | 30280 | 03981 |
| 47 | 09840 | 18332 | 23649 | 01547 | 49636 | 29561 | 02569 | 09900 | 29261 | 49581 |
| 48 | 43477 | 81376 | 77606 | 38759 | 39878 | 22936 | 84012 | 05418 | 58648 | 72566 |
| 49 | 82060 | 45252 | 15686 | 79666 | 70997 | 02076 | 10478 | 56938 | 83554 | 73624 |
| 50 | 63130 | 56495 | 13334 | 07633 | 00672 | 20053 | 05016 | 29832 | 23737 | 99474 |

**TABLE 4   Square Roots of Sample Sizes**

| Sample Size | Square Root of Sample Size | Sample Size | Square Root of Sample Size |
|---|---|---|---|
| 4 | 2.0 | 500 | 22.3 |
| 9 | 3.0 | 600 | 24.5 |
| 16 | 4.0 | 700 | 26.5 |
| 25 | 5.0 | 800 | 28.3 |
| 36 | 6.0 | 900 | 30.0 |
| 49 | 7.0 | 1000 | 31.6 |
| 64 | 8.0 | 1100 | 33.2 |
| 81 | 9.0 | 1200 | 34.6 |
| 100 | 10.0 | 1300 | 36.1 |
| 200 | 14.1 | 1500 | 38.7 |
| 300 | 17.3 | 2500 | 50.0 |
| 400 | 20.0 | 3600 | 60.0 |

**TABLE 5   Standardizing Values for the Wilcoxon Signed Rank Statistic**

| Sample Size | Mean | Standard Deviation |
|---|---|---|
| 10 | 27.5 | 9.8 |
| 11 | 33.0 | 11.2 |
| 12 | 39.0 | 12.7 |
| 13 | 45.5 | 14.3 |
| 14 | 52.5 | 15.9 |
| 15 | 60.0 | 17.6 |
| 16 | 68.0 | 19.3 |
| 17 | 76.5 | 21.1 |
| 18 | 85.5 | 23.0 |
| 19 | 95.0 | 24.8 |
| 20 | 105.0 | 26.8 |
| 21 | 115.5 | 28.8 |
| 22 | 126.5 | 30.8 |
| 23 | 138.0 | 32.9 |
| 24 | 150.0 | 35.0 |
| 25 | 162.5 | 37.2 |

**TABLE 6  Standardizing Values for the Wilcoxon Two-Sample Statistic**

Each cell gives the mean (top, marked *) and the standard deviation (bottom, marked **).

| | Smaller Sample Size | | | | | | | | | | | | | | | |
|---|---|---|---|---|---|---|---|---|---|---|---|---|---|---|---|---|
| | **5** | **6** | **7** | **8** | **9** | **10** | **11** | **12** | | | | | | | | |
| **5** | 27.5*<br>4.8** | | | | | | | | | | | | | | | |
| **6** | 36<br>5.5 | 39<br>6.2 | | | | | | | | | | | | | | |
| **7** | 45.5<br>6.2 | 49<br>7.0 | 52.5<br>7.8 | | | | | | | | | | | | | |
| **8** | 56<br>6.8 | 60<br>7.7 | 64<br>8.6 | 68<br>9.5 | | | | | | | | | | | | |
| **9** | 67.5<br>7.5 | 72<br>8.5 | 76.5<br>9.4 | 81<br>10.4 | 85.5<br>11.3 | | | | | | | | | | | |
| **10** | 80<br>8.2 | 85<br>9.3 | 90<br>10.2 | 95<br>11.3 | 100<br>12.2 | 105<br>13.2 | | | | | | | | | | |
| **11** | 93.5<br>8.8 | 99<br>9.9 | 104.5<br>11.0 | 110<br>12.1 | 115.5<br>13.2 | 121<br>14.2 | 126.5<br>15.2 | | | | | | | | | |
| **12** | 108<br>9.5 | 114<br>10.7 | 120<br>11.8 | 126<br>13.0 | 132<br>14.1 | 138<br>15.2 | 144<br>16.2 | 150<br>17.3 | | | | | | | | |
| **13** | 123.5<br>10.1 | 130<br>11.4 | 136.5<br>12.6 | 143<br>13.8 | 149.5<br>15.0 | 156<br>16.1 | 162.5<br>17.3 | 169<br>18.4 | 175.5<br>19.5 | | | | | | | |
| **14** | 140<br>10.8 | 147<br>12.1 | 154<br>13.4 | 161<br>14.7 | 168<br>15.9 | 175<br>17.1 | 182<br>18.3 | 189<br>19.4 | 196<br>20.6 | 203<br>21.8 | | | | | | |
| **15** | 157.5<br>11.5 | 165<br>12.8 | 172.5<br>14.2 | 180<br>15.5 | 187.5<br>16.8 | 195<br>18.0 | 202.5<br>19.3 | 210<br>20.5 | 217.5<br>21.7 | 225<br>22.9 | 232.5<br>24.1 | | | | | |
| **16** | 176<br>12.1 | 184<br>13.6 | 192<br>15.0 | 200<br>16.3 | 208<br>17.7 | 216<br>19.0 | 224<br>20.3 | 232<br>21.5 | 240<br>22.8 | 248<br>24.1 | 256<br>25.3 | 264<br>26.5 | | | | |
| **17** | 195.5<br>12.8 | 204<br>14.3 | 212.5<br>15.7 | 221<br>17.2 | 229.5<br>18.6 | 238<br>19.9 | 246.5<br>21.3 | 255<br>22.6 | 263.5<br>23.9 | 272<br>25.2 | 280.5<br>26.5 | 289<br>27.8 | 297.5<br>29.0 | | | |
| **18** | 216<br>13.4 | 225<br>15.0 | 234<br>16.5 | 243<br>18.0 | 252<br>19.4 | 261<br>20.9 | 270<br>22.2 | 279<br>23.6 | 288<br>25.0 | 297<br>26.3 | 306<br>27.7 | 315<br>29.0 | 324<br>30.3 | 333<br>31.6 | | |
| **19** | 237.5<br>14.1 | 247<br>15.7 | 256.5<br>17.3 | 266<br>18.8 | 275.5<br>20.3 | 285<br>21.8 | 294.5<br>23.2 | 304<br>24.7 | 313.5<br>26.1 | 323<br>27.5 | 332.5<br>28.8 | 342<br>30.2 | 351.5<br>31.6 | 361<br>32.9 | 370.5<br>34.2 | |
| **20** | 260<br>14.7 | 270<br>16.4 | 280<br>18.1 | 290<br>19.7 | 300<br>21.2 | 310<br>22.7 | 320<br>24.2 | 330<br>25.7 | 340<br>27.1 | 350<br>28.6 | 360<br>30.0 | 370<br>31.4 | 380<br>32.8 | 390<br>34.2 | 400<br>35.6 | 410<br>37.0 |

*mean
**standard deviation

# Appendix C: Standard Deviation

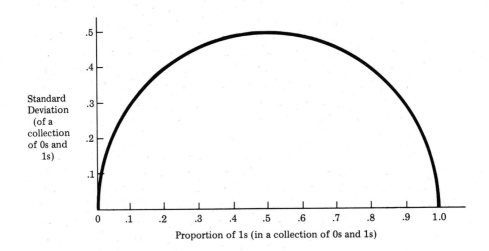

# Glossary

**Alternative Hypothesis** Complement, or opposite, of null hypothesis; assumed true if null hypothesis does not hold.

**Antecedent Factor** Agent preceding disease; may be a cause of the disease.

**Average** See indexes of central tendency.

**Bar Graph** A graphic display of categorical data; like a histogram but the "bars" are separated by spaces.

**Block** A group of homogeneous experimental units. (An analogous unit is a stratum of an existing population.)

**Categorical Data** See nominal data.

**Census** The study of an entire population of interest; a 100% sample.

**Chi-Square Distribution** A type of skewed (to the right) set of continuous data. Probabilities concerning such data are summarized in Appendix B, table 2.

**Cluster Sampling** A sampling procedure whereby clusters or groups of population elements, rather than individual elements, are selected.

**Confidence Interval** A 95% confidence interval is all values within two estimated standard errors of a sample proportion (mean).

**Continuous Data** Interval or ratio measurements that can theoretically vary by arbitrarily small amounts.

**Control Group** A group in an experiment or survey that is like the experimental group in all respects but does not receive a treatment.

**Convenience Sample** A sample selected in a convenient manner. No consideration is given to whether the sample is representative of the population of interest.

**Correlation** Association; the relationship between measurements.

**Correlation Coefficient** A statistic that measures the strength of a linear relationship between two measurements. We considered the Pearson product-moment correlation coefficient.

**Crime Rate** Number of occurrences of 7 major crimes reported per 100,000 population.

**Critical Region** See rejection region.

**Data**   See type of, continuous, discrete, ratio, interval, ordinal, or nominal.

**Datum**   A measurement taken on a population member (pl., data).

**Dependent Measurement**   The measurement we would like to predict.

**Descriptive Statistics**   Numbers used to describe certain characteristics of a collection of numbers.

**Dichotomous Population**   A collection of data that are categorized into one of two distinct categories. These two categories are labeled "1" and "0"; "1" denotes the category of interest.

**Discrete Data**   Interval or ratio measurements whose values vary by no more than a certain fixed amount; noncontinuous interval or ratio measurements.

**Distribution of the Sample Mean** (proportion)   A description of the change in the sample mean from (random) sample to (random) sample; also referred to as the sampling distribution of the mean.

**Double-Blind Experiment**   An experiment in which neither the technician nor the person being tested knows who received a treatment.

**Doublespeak**   "Inflated, involved, and often deliberately ambiguous language" (*Webster's New Collegiate Dictionary*, 1977).

**Empirical Rule**   Rule showing association between mean and standard deviation of a data set; applicable to unimodal symmetric data.

**Endemic**   The usual frequency of occurrence of a disease.

**Epidemic**   The unusually frequent occurrence of a disease.

**Epidemiology**   The study of the cause(s) of a disease, how a disease is transmitted in a population, and, ultimately, how the disease can be prevented.

**Estimate**   An estimate of a population parameter is an approximation of this unknown value. The sample statistic, which corresponds to this population parameter, is the approximation used.

**Event**   A collection of possible outcomes of an experiment.

**Experiment**   The study of a population that is created for the purpose of study.

**Experimental Group**   A group in an experiment that receives the treatment being tested.

**Experimental Unit**   Smallest unit receiving treatment in an experiment.

**Factor**   See type of, antecedent and outcome.

**Frame**  A complete description of a population. A frame could be a listing of all population elements. One might also describe a population using maps.

**Frequency Polygon**  Pictorial representation of the proportion of data that occurs in various areas of the data scale; characterized by lines connecting points representing data values and their frequency.

**Graphic Statistics**  A type of descriptive statistic device used to represent or summarize numerical information; a means of describing the distribution of numbers in a data set; pictorial representation of numbers.

**Histogram**  Pictorial representation of the proportion of data that occurs in various areas of the data scale; characterized by rectangles whose heights represent frequency and whose widths represent data values.

**Hypothesis**  See type of, null, alternative, or statistical.

**Independent Measurement**  A known measurement used to aid us in predicting the unknown level of the corresponding dependent measurement.

**Index Numbers**  Numbers that reduce data to values relative to the same quantity or base.

**Indexes of Central Tendency** (averages)  Statistics that describe the clustering tendencies of data. The mean, median, and modes are examples of this type of descriptive statistic.

**Indexes of Dispersion**  Statistics that describe how data tend to be spread. One usually describes dispersion about central values as the mean, the median, and the mode(s). The indexes of dispersion we discussed are the range, the standard deviation, and percentiles.

**Inferential Statistic**  Number derived from a sample that is used to infer to the corresponding population number (a parameter).

**Interval Data**  Data resulting from an actual measurement so that differences between data pairs are meaningful. These measurements do not have a meaningful zero.

**Judgment Sample**  A sample determined by an "expert" to be representative of a population.

**Least Squares**  A method of prediction whereby the sum of squared differences between actual values and predicted values in a sample are minimal.

**Level of Significance**  The probability of committing a Type-I error in a test of hypothesis.

**Linear Data Pairs**   The measurements under consideration have the property that large values of one measurement tend to be paired with large values of the other measurement. This is a positive relationship. If large values of one measurement are paired with small values of the other measurement, the relationship is negative.

**Mean**   The mean is the sum of all the data divided by the number of data. It is the "middle" of a data set in the sense that it is the center of gravity of the data.

**Median**   A number that has about half of the data in a data set larger than it and about half of the data smaller than it. You can think of the median as splitting an ordered array of data into two equal parts: all data in one part are larger than the median, and all the data in the other part are smaller than the median.

**Mode(s)**   A value(s) in a data set that occurs more frequently than surrounding data.

**Morbidity**   Data relating to occurrence of a disease.

**Mortality**   Data relating to death.

**Natural, or Census, Statistics**   Numbers that are collected through an investigation of an entire population.

**Nominal, or Categorical, Data**   Data sets characterized by each datum belonging to one and only one of certain categories or classes.

**Nonparametric Form**   A means of describing the distribution of data in a population in a general manner.

**Normal Distribution**   A data set that has a bell-shaped symmetric distribution and a unique relationship between the mean and standard deviation of the data. This relationship is summarized in Appendix B, table 1. (Recall that the percentages in table 1 can be thought of as probabilities when we think of the random experiment of randomly selecting a number from a normal data set.)

**Null Hypothesis**   The statistical hypothesis that is to be tested is called the null hypothesis.

**Observational Research**   The observation of events in their natural setting, being careful to do nothing that might influence the occurrence of the events of interest.

**Observational Unit**   The smallest element on which an observation is made in a sample survey.

**Ordinal Data**   Like categorical data with the exception that the data categories are ranked according to some criterion.

**Outcome Factor**   Whether or not a person has (had) a disease.

**Parameter**   A number that relates to a population. Examples of parameters would include indexes of central tendency and indexes of dispersion. Our discussions have been limited to the mean. This parameter would be the proportion of 1s when the population under investigation is a dichotomous population of 1s and 0s.

**Parametric Form**   A means of describing the distribution of data in a population in terms of parameters.

**Percentile**   A number that has a certain percentage of data in a data set below it. (Below in this case means smaller in value.) Certain percentiles describe the dispersion of data. This method of indicating dispersion is most effective for data sets in which the amount of dispersion varies in different areas of the data scale.

**Placebo**   A substance like a particular treatment in appearance but inactive as far as the response of interest is concerned.

**Point-in-Time Survey**   Taking a sample from a population as the population exists at a particular time.

**Population**   The entire collection on which we desire information; can be thought of as a collection of numbers.

**Poststratification**   See stratification.

**Probability** (relative frequency)   The probability of an event is the proportion of times that the event occurs when a random experiment is run a large number of times.

**Prospective Study**   The study of a group in a continuing manner over an extended period of time; people with and without a particular antecedent factor are selected with the proportions getting the disease being compared.

**Quota Sample**   A type of prestratified sample in which elements are selected from a stratum by nonrandom means.

**Random Experiment**   An experiment for which all possible outcomes are known but the outcome of a particular execution of the experiment cannot be predicted with certainty. In statistical inference we consider the selection of a random sample from a population as the random experiment of interest. (See chapter 5 for a discussion of how a random sample might be selected as well as how this random experiment allows us to measure [using probability] the uncertainty of statistical inference.)

**Randomization**   Assignment of treatment to experimental units according to an outside random mechanism.

**Range**   Refers to either the pair of data that are the largest and smallest values in a data set or the number that is the difference between the largest and smallest values.

**Rank Tests** Tests based on the ranks of data or on the ranks of the deviations of data from specified values.

**Ratio Data** Interval data with a meaningful zero.

**Real, or Constant, Dollars** Monetary amounts reflecting a fixed buying power for the dollar.

**Rejection, or Critical, Region** The values of the sample statistic that will lead to rejection of the null hypothesis.

**Retrospective Survey** A backward look for the cause of a disease that involves people who already have the disease in question.

**Sample** A part of a population that is studied in order to derive some insight into unknown population characteristics. Also, *see* type of, simple random, systematic, stratified random, quota, judgment, convenience, or cluster.

**Sample Allocation** A determination of what part of a sample will be selected from each strata in a prestratified sample.

**Sample Survey** The study of an existing population. A 100% sample is called a *census*.

**Sampling Distribution** *See* distribution of the sample mean.

**Sampling Unit** The smallest element used in the selection process of a sample survey.

**Scatter Diagram** Graphic representation of data pairs.

**Semi-Log Graph** Graph that depicts rates of change as opposed to actual arithmetic changes as in an arithmetic graph.

**Sign Test** Test based on the sign, plus or minus, of the deviations of data from specified values.

**Simple Random Sample** A sample drawn from a population is a simple random sample if all samples of the same size have the same chance of being the sample.

**Standard Deviation** The (positive) square root of the variance. The empirical rule describes the dispersion of data about the mean in units of standard deviations. This rule is most accurate for data that are unimodal and symmetric.

**Standard Error of a Statistic** The standard deviation of the sampling distribution of a statistic.

**Statistical Doublespeak** A type of doublespeak that comes from the use of a meaningless number or abuse of a credible statistic. This type of doublespeak can be avoided if meaningless numbers can be detected and statistics are properly interpreted.

**Statistical Hypothesis** A statement about a parameter of a population. In the case of a dichotomous population, a statistical hypothesis is a statement about the proportion of 1s in the population.

**Statistic**   A number; see type of, descriptive, inferential, natural, vital, and graphic.

**Statistics**   The design of an experiment or sample survey and the collection, analysis, and interpretation of the resulting data. Statistics can also be thought of as a language. It is the language of statistics that is discussed in this book.

**Stratification**   The division of a collection of data into parts. Prestratification indicates that the population is stratified with sampling being carried out within each of the stratum. This helps ensure that each stratum is properly represented in the overall sample. Poststratification refers to dividing a sample into parts. We would often like to be sure that the sample does represent certain groups. The sample is broken into strata when prestratification is not economically or practically possible.

**Stratified Random Sample**   A prestratified sample in which the elements selected from each stratum form a simple random sampling of the stratum elements.

**Survey**   See type of, point-in-time, prospective, and retrospective.

**Systematic Sampling**   A sample selected by starting at a random element in a population and taking every so many population elements.

**Target Population**   The population that is to be studied by a sample survey.

**Type-I Error**   The error of rejecting a true null hypothesis.

**Type-II Error**   The error of failing to reject a false null hypothesis.

**Variance**   The average squared deviation of data from the mean.

**Vital Statistics**   Data relating to birth, death (mortality), marriage, divorce, and the occurrence of disease (morbidity).

# Index

Public doublespeak
Statistical hypothesis: 160–171; level of significance, 165; null hypothesis, 163–166, 168; rejection region, 166; tests, 161–171. *See also* Population mean
Statistical inference: 86–87; definition, 8–9, 114; Ronald Fisher, 8–9. *See also* Background influences on statistics; Population; Probability; Sample proportion; Sample survey; Sampling distribution of the mean; Sampling procedures; Statistical hypothesis
Statistical interpretation: 19–36. *See also* Background influences on statistics
Statistician, job description: 1, 5
Statistics: general objectives, 1–3; language, 5. *See also* Descriptive statistics; History of statistics; Interpretation; Statistical doublespeak; Statistical inference
Stratification. *See* Prestratification
Stratified random sample: 123, 126
Stratified sampling. *See* Prestratification
Stratum: 123
Survey. *See* Sample survey

Survey design. *See* Sample survey design
Symmetry: 49; point of, 96
Systematic sampling: 129–130

Table of random digits. *See* Random digit table
Target population: 223–224
Type-I Error: 164–165
Type-II Error: 164–165

Uncertainty. *See* Probability
Utility data: 75, 77

Variance: 69, 71, 77–78, 169. *See also* Standard deviation
Vital statistics: 198

Wald, Abraham: 14
Wholesale Price Index: 52–53, 287–288
Wilcoxon rank-sum test: 180–181; ranking, 181–182; standardizing, 182, 309
Wilcoxon signed-rank test: 179–181; standardizing, 180, 308

Yankelovich, Daniel: 30